European Federation of Corrosion
Publications
NUMBER 1

A Working Party Report

on Corrosion in the Nuclear Industry

Published for the European Federation of Corrosion

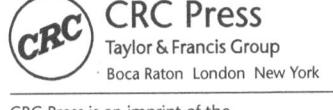

CRC Press
Taylor & Francis Group
Boca Raton London New York

CRC Press is an imprint of the
Taylor & Francis Group, an **informa** business

First published 1989 by The Institute of Metals

Published 2019 by CRC Press
Taylor & Francis Group
6000 Broken Sound Parkway NW, Suite 300
Boca Raton, FL 33487-2742

ISBN 13: 978-0-901462-73-2 (pbk)
ISBN 13: 978-1-138-44147-7 (hbk)

**Visit the Taylor & Francis Web site at
http://www.taylorandfrancis.com**

**and the CRC Press Web site at
http://www.crcpress.com**

Compiled by the Institute's CRC unit from original typescripts
and illustrations provided by the authors

British Library Cataloguing in Publication Data
Corrosion in the nuclear industry.
1. Nuclear power stations. Equipment.
Corrosion
I. European Federation
of Corrosion II. Series
621.48'33

Book Number 481

Contents

European Federation of Corrosion
- Series Introduction

The EFC, incorporated in Belgium, was founded in 1955 with the purpose of promoting European co-operation in the fields of research into corrosion and corrosion prevention.

Membership is based upon participation by corrosion societies and committees in technical Working Parties. Member societies appoint delegates to Working Parties, whose membership is also expanded by co-option of other individuals. The activities of the Working Parties cover corrosion topics associated with inhibition, education, reinforcement in concrete, microbial effects, hot gases and combustion products, environment sensitive fracture, marine environments, surface science, physico-chemical methods of measurement, the nuclear industry, and computer based information systems. Working Parties on other topics are established as required.

The Working Parties function in various ways, e.g. by preparing reports, organising symposia, conducting intensive courses, and producing instructional material, including films.

The activities of the Working Parties are co-ordinated, through a Science and Technology Advisory Committee, by the Scientific Secretary.

The administration of the EFC is handled by three Secretariats: Dechema in the Federal Republic of Germany, the Societe de Chimie Industrielle in France, and the Institute of Metals in the United Kingdom. These three Secretariats meet at the Board of Administrators of the EFC.

There is an annual General Assembly at which delegates from all member societies meet to determine and approve EFC policy.

News of EFC activities, forthcoming conferences, courses etc. is published in a range of accredited corrosion and certain other journals throughout Europe. More detailed descriptions of activities are given in an occasional Newsletter prepared by the Scientific Secretary.

The output of the EFC takes various forms. Papers on particular topics, for example, reviews or results of experimental work, may be published in scientific and technical journals in one or more countries in Europe; Conference proceedings are often published by the organisation responsible for the conference.

In 1987, the Institute of Metals was appointed as the official EFC publisher. Although the arrangement is non-exclusive and other routes for publication are still available, it is expected that the Working Parties of the EFC will use the Institute of Metals for publication of reports, proceedings etc. wherever possible.

A D Mercer
Scientific Secretary at the Institute of Metals
London, UK

EFC Secretariats are located at:

Robert Wood
European Federation of Corrosion
The Institute of Metals
1 Carlton House Terrace
LONDON SW1 5DB, UK

Dr D Behrens
Europaische Foderation Korrosion
DECHEMA
Theodor-Heuss-Allee 25
D-6000
FRANKFURT (M)
F R G

M R Mas
Federation Europeene de la Corrosion
Societe de Chimie Industrielle
28 Rue Saint-Dominique
F-75007 PARIS
FRANCE

Introduction

In 1986 the EFC Working Party on Nuclear Corrosion was reorganised with the objective of concentrating on nine topics of relevance to the nuclear power industry. The group of experts in the Working Party is dedicated to collecting information on corrosion in this industry, to the analysis of the data obtained, and to transferring the information to scientists and engineers in the industry. Successes in overcoming problems and the need for further research with also form part of the activities of the Working Party.

The new structure of the Working Party is based on the following topics:

- Pressurised Water Reactors
- Boiling Water Reactors
- Fuel Elements (Cladding)
- Advanced Gas Reactors
- High Temperature Reactors
- Liquid Metal Fast Breeders
- Fusion Reactors
- Reprocessing
- Waste Management (Disposal)

The first meeting of the Working Party following the restructuring was on the occasion of EUROCORR '87 in Frankfurt. Reports were presented by experts on the major corrosion problems in a number of these special topics, i.e. a 'state-of-the-art' review was made with emphasis on unresolved problems needing additional research and development.

These reports were available as preprints to those attending EUROCORR '87 where they were well-received. It was then generally recognised that the information should be made available to a wider readership through a formal publication.

The present volume has therefore been prepared and represents Number One in the series of EFC publications.

Ph. Berge
Chairman of the Working Party

1 Corrosion in Pressurized Water Reactors

J. -Ph. Berge

Electricité de France, Service de la Production
Thermique, Groupe des Laboratoires
Carrefour Pleyel - 21, Allée Privée
F-93206 Saint-Denis Cedex 01

For many years now, in response to the multiple types of corrosion-induced damage encountered in PWR components, a major R & D effort has been underway to determine the best materials, fabrication processes, and operating conditions with which to combat corrosion phenomena.

From the plant operator's viewpoint, corrosion is a source of costly inspections, repairs, and outages. Above all, it poses a potential threat to reactor safety, particularly when the reactor coolant pressure boundary is affected.

This paper does not address fuel cladding corrosion problems, covered by Dr. Leistikow, or conventional island components. Though corrosion damage has been detected in condenser tubes, turbine rotors and nozzle rings, generator binding bands, and steam/feedwater systems, these cases are not specifically "nuclear" in nature and are covered by other groups in the European Federation of Corrosion.

For the purposes of this paper, corrosion effects will be classified according to the three groups of materials affected :

- austenitic stainless steels (piping and reactor internals) ;

- heat exchanger materials (steam generators in particular) ;

- high-strength materials (bolting, hardfacing surfaces).

For each group of materials, we shall present examples of corrosion damage, R & D efforts and resultant improvements, and remaining problems with respect to plant reliability and safety.

Finally, we shall discuss in a fourth part of this presentation corrosion products conveyed through the reactor coolant system - whose consequences on operating and maintenance personnel dose rates are significant.

STAINLESS STEELS IN REACTOR COOLANT AND AUXILIARY SYSTEMS AND REACTOR INTERNALS

Austenitic stainless steels -the most widely used materials in these systems- have experienced two types of problems :

- Incidents resulting from accidental contamination, particularly by halogenated products. Stress corrosion has been observed in the presence of chlorides produced by high-temperature decomposition of synthetic products. The solution lies in tighter cleanness control and stringent specifications for solvents, paint, grease, etc.

- Concentrated boric acid, even at room temperature, can attack stainless steel pipe in the presence of oxygen. Welds that are sensitized to intergranular corrosion are particularly susceptible to this form of deterioration.

A further type of corrosion observed recently and resulting in mostly transgranular cracks is currently under investigation in the laboratory. This form of attack occurs in the presence of concentrated boric acid at temperatures of 100 to 300° C, even though no other impurities, such as halogenated compound, are present.

This phenomenon has been found to affect stainless steels, high-nickel alloys and high-strength alloys.

Further research, however, will be required to determine the exact conditions associated with this type of corrosion and to understand the mechanism involved. A relationship, however, has already been shown to exist between the combined presence of boric acid and chlorides and the occurence of cracking.

STEAM GENERATORS AND OTHER HEAT EXCHANGERS

In terms of the diversity of types of damage, the amount of applied research that it has generated, the debate that has surrounded the results obtained since 1959, and, more recently, its practical repercussions corrosion of steam generator tubes poses problems on a scale rarely seen in an industrial context.

	C	S	P	Si	Mn	Ni	Cr	Mo	Ti	Al	Fe	Cu	Co
Alloy 600 ASTM B 163	0.15	0.015	---	0.5	1.0	Balance (72)	14 to 17	---	---	---	6 to 10	0.5	0.10
Specification for EDF	0.010 to 0.050	0.015	0.025	0.50	1.00	7200	1400 to 1200 1550 a	---	0.50	0.50	6.00 to 10.00	0.50	0.10 0.05 b
Alloy 690 ASMR Code case 14.84.3	0.05	0.015	---	0.50	0.50	58	27 to 31	---	---	---	7 to 11	0.50	0.10
Alloy 800 ASTM B 163	0.10	0.015	---	1.0	1.5	30 to 35	19 to 23	---	0.15 to 0.6	0.15 to 0.6	balance	0.75	0.10
Type 316 Mainless steel ASTM A 376	0.08	0.03	0.03	0.75	2.0	11 to 14	16 to 18	2 to 3	---	---	balance	---	0.10

a - Desirable minimum value
b - Desirable maximum value

The table above shows the materials of construction used by various vendors. The mechanical properties, thermal characteristics and chemical composition of these materials are described elsewhere (Ph. Berge - J.R. Donati - Nuclear Technology - Vol. 55 Oct.81). In response to potential corrosion risks or to actual corrosion problems, new materials and fabrication processes have been adopted and system operating parameters adjusted accordingly, e.g. through water chemistry control. At the same time, new causative agents have been identified.

- Secondary side stress corrosion cracking, attributable typically to a strongly alkaline environment resulting from boiling of slightly contaminated water in crevice locations, or to decomposition of trisodium phosphate used for secondary water treatment.

- Localized chemical attack from acid phosphate residues (wastage).

- Buildup of magnetite in tube-to-tube support plate annuli (denting) in cases where phosphate treatment has been discontinued to resolve previous problems, giving rise to concentrations of unbuffered acidic chloride solutions in crevices.

- Primary side stress corrosion cracking of the most widely used material, Inconel Alloy 600 containing 70 % nickel. This phenomenon was reported in 1959 and was a subject of debate for almost 20 years before the first serious consequences -cracking in highly stressed regions of Alloy 600 tubes- were observed.

Most secondary side and support plate corrosion problems can be resolved by stringent waterchemistry control.

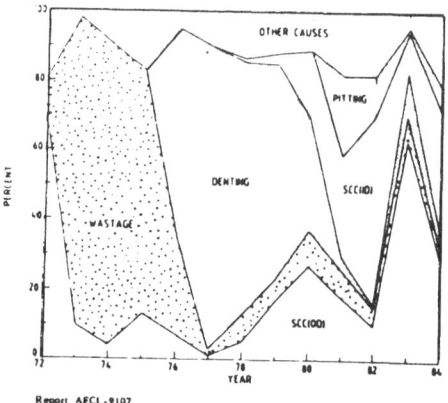

Report AECL-9107

However, water chemistry is generally not a significant factor in primary side cracking of structurally susceptible Alloy 600. To eliminate this type of degradation and determine the presence or extent of damage, a number of measures have been implemented :

- In plants currently under construction, Alloy 600 has been replaced by Alloy 690 or the earlier Alloy 800. Advantages and disadvantages of these materials have been discussed elsewhere.

- In plants built over the past ten years, the structure of Alloy 600 was improved and residual stresses from bending and straightening operations reduced by a final vacuum heat treatment for 15 hours at about 700°C.

- In operating plants that experienced cracking in regions subject to high levels of residual stress, such as small-radius bends and expanded tube-to-tubesheet joints, the former have undergone stress relieving heat treatment, while the latter have received a prestressing treatment by shot peening.

Although the large number of cracked tubes in a single component might seem to constitute a threat to plant safety, examination of tube samples and laboratory tests have shown that cracks in the roll transition at the upper surface of the tubesheet are axially oriented. It has also been demonstrated that,

2

even in the event of secondary side depressurization, these cracks will result in primary-to-secondary leakage before growing to a size likely to cause a significant tube failure.

Analyses are performed to determine exceptional cases, where this "leak before break" criterion might not apply because of fabrication anomalies, and tubes are plugged accordingly as a preventive measure.

To conclude, it seems probable that, thanks to R & D programs, future reactors will be spared the above corrosion problems, provided that rigorous water chemistry control is maintained and that the necessary material and design modifications are introduced.

As far as our understanding of the various corrosion mechanisms is concerned, tests in sodium hydroxide solutions and pure water have demonstrated the significance of temperature and metallurgical structure. However, there are variations in behavior between the different alloys that are not yet understood, and further investigation, particularly with respect to grain boundary creep, will be required to explain the effects observed and predict the behavior of other alloys.

BOLTING MATERIALS

Various types of bolting material corrosion have been observed under service conditions.

- Corrosion of valve and seal assembly bolting due to boric acid buildup following reactor coolant leakage.

  Under these conditions, low alloyed steel bolting is heavily corroded by boric acid, while stainless steels resist generalized corrosion. But a comprehensive study of the various materials with an austenitic matrix showed almost all of them to be susceptible to stress corrosion cracking at 300-350° C in a 40 % solution of boric acid. This phenomenon is very similar to the stress corrosion cracking of stainless steels in concentrated boric acid discussed above.

  Here again, the corrosion mechanism is not yet fully understood and, in the absence of a reliable solution, the condition of bolting materials must be monitored.

- Primary fluid stress corrosion cracking of Alloy X 750 guide tube support pins and bolts :

  Numerous failures due to this type of corrosion -predicted by laboratory tests in the 1960s- have been reported. The mechanisms involved appear to be similar to those resulting in cracking of Alloy 600. In this case, too, the solution lay in modifying the metal structure by an appropriate heat treatment resulting in fine precipitation of chromium carbide at grain boundaries to improve corrosion resistance in pure water and sodium hydroxide, and in reducing stress levels by modifying part geometry or tightening torque.

HARD FACING ON RUBBING SURFACES (VALVES)

Deterioration of hard facing alloys has also been observed in some cases. This type of corrosion can result not only in loss of leaktightness or binding (e.g. in valves), but also in elevated dose rates from cobalt-60 released into the reactor coolant system, since most hard facing alloys are cobalt-based. During pickling or exposure to borated water, corrosion can occur in zones depleted in chromium through excessive carburizing when the hard facing was deposited.

CORROSION PRODUCTS

A considerable research effort is currently being devoted to determining optimum reactor coolant chemistry (injection of lithium hydroxide) when the concentration of boric acid is adjusted to compensate for core reactivity variations. A number of measures have been considered to minimize the quantity of oxides released by system components and to avoid transport of corrosion products, which can become activated by contact with fuel elements and accumulate at various locations in the system. These include maintaining constant pH, selection of optimum pH, and end-of-cycle pH adjustments. Water chemistry specification must also consider the risk of damage to the fuel cladding, in the event of local boiling, and to other system components.

An understanding of transfer mechanisms, painstaking laboratory analyses, and comparison of detailed reports from operating plants should indicate the best water chemistry specifications to deal with this problem. In addition, the elimination of cobalt alloys and a reduction of average trace values of cobalt in structural materials should have a significant impact on corrosion product activation problems.

CONCLUSION

Over the 25 years since the first PWR nuclear power plants came into service, numerous corrosion problems have been encountered in various components. Solutions to date have relied on an increased understanding of the mechanisms involved, adherence to good practice in material selection and fabrication, and stringent compliance with water chemistry specifications. Certain operators are still paying now for the failure to observe these principles in the past. In this domain, as in many others, numerous problems could have been avoided by closer collaboration between research laboratories and plant constructors or operators.

2 Corrosion Problems in Boiling Water Reactors and their Remedies

B. Rosborg

The author is Manager of the Department for Materials Technology at Studsvik AB, Sweden.

SYNOPSIS

This article briefly presents current corrosion problems in boiling water reactors and their remedies. The problems are different forms of environmentally assisted cracking, and the remedies are divided into material-, environment-, and stress-related remedies.

INTRODUCTION

This presentation of corrosion problems in boiling water reactors (BWRs) and their remedies is mainly limited to current corrosion problems in the nuclear steam supply system of direct-cycle BWRs, excluding Zircaloy fuel cladding behaviour which is the subject of a separate presentation. The corrosion problems amount to a list of environmentally assisted cracking problems, but for the general and significant maintenance problem associated with contamination due to radioactive corrosion products (1-5). The list comprises

intergranular stress corrosion cracking (IGSCC) in weld-sensitized stainless steel piping

IGSCC in cold-bent stainless steel piping

irradiation-assisted stress corrosion cracking (IASCC) in stainless alloys

IGSCC in high-strength stainless alloys

A prospective corrosion problem as judged from literature references and one which relates to plant life, is

corrosion fatigue in pressure vessel steel

since the reactor pressure vessel is the most critical component in the BWR pressure boundary as regards plant safety.

The current corrosion problems are not directly related to safety. However, they affect the overall plant performance and availability. Past experience has demonstrated the need for better understanding of materials degradation phenomena related to corrosion and irradiation in order to find remedies leading to improved plant productivity.

IGSCC IN WELD-SENSITIZED STAINLESS STEEL PIPING

A great number of incidents of IGSCC in the weld heat-affected zones of Type 304 stainless steel piping have occurred since 1974 (3). During the period 1974 - 1984 more than 650 cases of IGSCC were found. These incidents, together with the earlier incidents of IGSCC in furnace-sensitized pressure vessel nozzle safe-ends of Type 304 and 316 stainless steel, represent the single largest source of productivity loss related to the BWR pressure boundary. However, in spite of this only a very low percentage of the many thousands of welds in a BWR has experienced cracking. The greatest frequency of cracking incidents has been reported in recirculation piping systems.

The conditions for cracking are a weld-sensitized (or furnace-sensitized) microstructure, oxygenated water, and high tensile stresses.

Several remedies are available for handling this corrosion problem (3)(6-7). As regards material-related parameters the carbon content of the stainless steel and the welding procedures, particularly the heat input, are important. The most important environmental parameter is the corrosion potential of the steel as cracking can be avoided by keeping the corrosion potential below a certain critical potential for IGSCC. However, this potential is influenced by the impurity content of the reactor coolant. As regards tensile stresses the weld residual stresses usually represent a substantial part.

Material-related remedies - An obvious material-related remedy is to replace the susceptible material with a more resistant material. Alternative materials are Types 304 and 316 Nuclear Grade stainless steels with max 0.02 % C and max 0.10 % N. By this low carbon content weld sensitization is avoided, while the nitrogen provides for design allowable stresses similar to that of conventional stainless steel grades. The Nuclear Grade stainless steels are used combined with improved welding procedures which limit the heat input. Type 347 stainless steel, which has successfully been used in West Germany, and carbon steel are additional alternative materials. Several recirculation piping replacements have been carried out since 1982. However, they are costly.

Another obvious material-related remedy is to use solution heat treatment to eliminate sensitization and relieve the weld residual stresses.

Corrosion-resistant cladding is a remedy which has find use for plants under construction. Both shop and field procedures are available.

Environment-related remedies - Hydrogen dosage to the feedwater in order to suppress the oxygen and hydrogen peroxide content of the reactor coolant, and thus lowering the corrosion potential of the material, is an interesting remedy for operating plants. The first hydrogen water chemistry test in a commercial reactor was performed in Sweden in 1979. As impurities in the reactor coolant affect the susceptibility to cracking, hydrogen water chemistry is applied together with stringent water quality control. Hydrogen water chemistry has been adopted for evaluation in both Sweden and the US.

Stress-related remedies - Several stress-related remedies are available. In induction heating stress improvement the outer surface of the pipe is heated to about 550°C while the inner surface is cooled with water leaving compressive residual stresses on the inside. It was first used in Japan in 1977, and has since then been used extensively. Several hundreds of weldments have been treated.

Pipelocks are mechanical devices which are applied to hold cracked weldments together and induce favourable compressive stresses on the inside of the pipe. In mechanical stress improvement a pipelock device is assembled on the pipe, tightened to induce compressive stresses on the inside, and then removed.

In heat-sink welding, which is a remedy for plants under construction, the inside surface of the pipe is cooled with water during welding leaving a favourable residual stress pattern on the inside. In last pass heat-sink welding inside cooling is only applied during the last welding pass with a high heat input.

In heat-sink rewelding the outer surface of the weldment is ground off and then refilled by welding while the inner surface is cooled with water.

Interim remedies - Weld overlay (weld buttering) and outside sleeve repairs and so-called flawed pipe analysis methodology have found use as interim remedies. In weld-overlay repairs Type 308L is overlay welded on the outside of the pipe while the inside is cooled with the reactor water. A weld overlay of the same thickness as the original pipe may be used, or a so-called mini-overlay design may be used. Application of an outside sleeve is a method where a sleeve, split in two halves, is placed on the outer surface of the pipe and welded to it.

In a flawed-pipe analysis the remaining life of the pipe is determined according to code requirements from the knowledge of stress levels, environmental effects, and crack-growth data. This methodology is mainly of interest for large diameter pipes.

Design-related remedy - One design-related remedy is the use of internal recirculation pumps, thus avoiding external recirculation lines.

IGSCC IN COLD-BENT STAINLESS STEEL PIPING

IGSCC in cold-bent Type 304 stainless steel piping has been found in only a few reactors in Sweden and in the US. The cracking is not related to weld-sensitization but to cold-bending. While longitudinal cracks were found in the earlier failures, recently circumferential cracks have also been found.

Material-related remedies - Even if the understanding of this cracking is limited, annealing after cold-bending of Type 304 or the use of Type 316 instead of Type 304 are possible remedies.

IASCC IN STAINLESS ALLOYS

IASCC is a type of IGSCC which has caused failures in reactor internal components made of austenitic stainless alloys of various kinds, including Type 304, 316, 321 and 348 stainless steels, Incoloy 800, and Inconel 600, 625, 718 and X-750 (5). It was first observed for stainless steel fuel cladding.

The cracking occurs in apparently non-sensitized material, and failures have been observed not only in BWRs but also in PWRs. The BWR environment seems, however, more accelerating. A fluence threshold seems to exist.

Even if the failures so far have occurred in easily replaceable parts, there is a major concern over the long-term performance of not easily replaceable parts.

5

Material-related remedies - Materials low in P and Si seem to have high resistance to cracking.

Environment-related remedy - Laboratory testing has shown that hydrogen water chemistry may be favourable.

IGSCC IN HIGH-STRENGTH STAINLESS ALLOYS

The age-hardenable austenitic stainless alloys Inconel X-750 and A-286 are used in BWR internals, such as beams, bolts, screws and springs. Several different heat treatments are used for Inconel X-750.

Extensive cracking has occurred in both Inconel X-750 and A-286 components (8). In some cases only a small fraction of the components has been affected. In other a substantial part has been affected. Some failures have not caused any problems for plant operation, while others have caused shutdowns. Several of the different heat treatments for Inconel X-750 have revealed IGSCC. The most vulnerable heat treatment has been the so-called "equalized and aged" treatment. The material behaviour seems to be similar in both BWR and PWR environments.

Material-related remedies - At least two material-related remedies have been used, that is either change of material or use of improved heat treatments. Thus, A-286 has been exchanged for either Type 304L and 316L stainless steel or low alloy steel. As to Inconel X-750 a high temperature anneal (1090°C) followed by a single aging treatment at 705°C for 20 h has been proposed.

Stress-related remedies - Use of lower design stresses has been effective in mitigating cracking. Satisfactory behaviour has been shown at moderate and low stress levels.

CORROSION FATIGUE IN PRESSURE VESSEL STEEL

It has been stated that corrosion fatigue appears to be the primary subcritical crack growth mechanism in pressure vessel steel (9). However, no current corrosion problems as to corrosion fatigue in pressure vessel steel exist. Earlier incidents of BWR feedwater nozzle and control-rod-drive return line nozzle cracking have been reported. These were, however, attributed to initiation of cracks in the stainless steel cladding due to thermal fatigue, which then propagated a small distance into the pressure vessel steel until they were discovered and ground away.

The environment has a major influence on corrosion fatigue in pressure vessel steel at lower frequencies. However, the role of oxygen in the reactor coolant is not yet clear, since high crack growth rates can be obtained in both BWR and PWR simulated environments, at least for steel with high sulphur contents.

Material-related remedies - Steels with low sulphur contents (< 0.008 % S), and appropriate sulphide morphologies, should be used for new vessels.

Stress-related remedies - The corrosion fatigue crack growth rate is strongly dependent upon loading variables such as the stress-intensity range, the mean load, and the frequency. Engineering codes, for example the ASME Boiler and Pressure Vessel Code, provide means of evaluating any defect revealed by in-service inspection. Within an International Cooperative Group on Cyclic Crack Growth Rate a database has been established, see below, which proposes new reference curves for the ASME Code as to corrosion fatigue in pressure vessel steel.

EPRI DATABASE ON ENVIRONMENTALLY ASSISTED CRACKING

The predominant materials degradation mechanism in operating nuclear power plants is environmentally assisted cracking. In order to support the development of models to predict crack growth in reactor materials and environments, and to assist in the development of engineering codes for reactor pressure vessel and piping steels, the Electric Power Research Institute (EPRI) compiles a Database for Environmentally Assisted Cracking (EDEAC)(10). EDEAC is meant to be a source of all available measurements of crack growth rates in various material-environment combinations found in nuclear power plants. It contains data on more than 3400 crack growth tests on reactor materials.

REFERENCES

1 ROBERTS, J T A
   Structural Material in Nuclear Power Systems.
   Plenum, New York, 1981.

2 NORRING, K and ROSBORG, B
   A compilation of experiences of corrosion in Nordic nuclear power plants.
   Studsvik AB, 1985 (STUDSVIK/EI-85/43).

3 DANKO, J C
   Boiling water reactor research on pipe cracking.
   Materials Performance 24 (1985):5 p 14-17.

4 HÄNNINEN, H und AHO-MANTILA, I
   Umgebungsinduzierte Rissbildung bei Werkstoffen in druckführenden Bauteilen von Leichtwasserreaktoren.
   Der Maschinenschaden 59 (1986):4 s 154-164.

5 Effects of Irradiation on Stress Corrosion Cracking.
   Results of the Research Assistance Task Force Meeting, held June 3 and 4, 1986. EPRI NDE Center, Charlotte, North Carolina, November 1986.

6 EPRI 1986 Seminar on Countermeasures for Pipe Cracking in BWRs.
   Palo Alto, California, November 1986.

7 EPRI 1986 Seminar on BWR Corrosion, Chemistry, and Radiation Control.
   Palo Alto, California, November 1986.

8 McILREE, A R
   Degradation of high strength austenitic alloys X-750, 718, and A-286 in nuclear power systems.
   Proc Inter Symp on Environmental Degradation of Materials in Nuclear Power Systems - Water Reactors, NACE, Houston 1984, p 838-850.

9 Proc Second IAEA Specialists' Meeting on
   Subcritical Crack Growth.
   NUREG/CP-0067, 1986.

10 EPRI Database for Environmentally Assisted
   Cracking (EDEAC).
   Electric Power Research Institute, 1986
   (EPRI NP-4485).

3 Zircaloy Fuel Cladding Corrosion Behaviour under Light Water Reactor Operation and Accident Conditions

S. Leistikow

Prof. Dr. Leistikow is in the Department of the
Materials and Solid State Research Institute II
of the Nuclear Research Center Karlsruhe

SYNOPSIS

A main subject to be considered when an extension
of the burnup of Light Water Reactor fuel is
discussed, is the water/steam corrosion of the
Zircaloy fuel cladding material under reactor
operation and hypothetical accident conditions.
    Here it is pointed out that Zircaloy
corrosion under normal reactor operation
conditions increases as function of burnup at
least linearly. Relying on the function of the
cladding wall as a first barrier against fuel and
fission product release to the environment a
suitable rod design should take into account -
besides corrosion under normal LWR operation
conditions - an additional loss of wall thickness
by steam oxidation in case of a hypothetical
loss-of-coolant accident. This aspect has been
verified by high temperature measurements of
Zircaloy oxidation kinetics and creep-rupture
behavior under isothermal and temperature-
transient conditions in steam.

1. CORROSION UNDER NORMAL LWR OPERATION CONDITIONS

The corrosion behavior of Zircaloy fuel cladding
in Light Water Reactors (LWR) was followed up
from the beginning of their in-pile application.
Initially, the corrosion-related hydrogen take-up
and embrittlement were considered to be serious
lifetime limiting effects, but meanwhile this
assumption has been disproved. Today, corrosion
leading to excessive growth of the oxide scale
and the risk of oxide spalling, besides its
interrelation to crud deposition /1/, are of
major concern for the anticipated extended fuel
exposure.

1.1 EX-REACTOR WATER CORROSION

Generally, the corrosion reaction between
Zircaloy and pure high temperature water or steam
can be expressed by the equation: $Zr + 2H_2O \rightarrow
ZrO_2 + 2H_2$. Part of the corrosion product
hydrogen diffuses through the oxide layer into
the metal. Its amount, expressed as a percentage
of the total amount generated by metal corrosion,
is called "pick-up fraction" /2/. While the
corrosion kinetics of the different Zircaloys are
similar, the pick-up fractions of Zircaloy-4 are
smaller than for Zircaloy-2.
    The initial corrosion kinetics of Zircaloy
in water or steam in the temperature range 250-
400°C can be approximated by a cubic rate law
(Fig. 1). Under these conditions the uniform
oxide layer normally formed is smooth, continuous
black or grey-black lustrous, and adherent. It is
protective in nature. At a weight gain of
approximately 30-40 mg/dm^2, corresponding to 2-
2.7 μm oxide layer thickness, the corrosion
kinetics turn over to follow a linear rate law
while the corrosion product remains black. After
extensive exposure, the film may become mottled,
then grey, and finally tan, still retaining its
adherence to the underlying metal.
    In contrast to the normally uniform
corrosion, a localized corrosion attack has been
observed on Zircaloy specimens after exposure to
high pressure steam at temperatures $\geq$ 475°C. The
formation of local oxide lenses (nodules or
pustules) is called nodular corrosion. The

patches form in an otherwise uniform-appearing corrosion layer and reach locally a much larger thickness than the uniform oxide.

2. IN-REACTOR WATER CORROSION

The in-reactor corrosion behavior of Zircaloy is different from the ex-reactor behavior. Oxygen content of the primary coolant is one factor which markedly influences the in-pile behavior. Under oxygenated system conditions, fast neutron flux is a controlling factor whereas under reduced oxygen levels temperature may become more important. Other factors which may affect in-reactor corrosion include fuel rod power, coolant chemistry and pH, mass flow, crud, and prior oxidation history.

2.1 BOILING WATER REACTOR (BWR) SYSTEMS

BWR exposure conditions are
- 280-300°C,
- 70 bar system pressure,
- 0.2 ppm O_2 (formed by radiolysis),
- high purity water, no additives,
- fast neutron flux $5 \cdot 10^{13} n/cm^2 s$.

Under these conditions, enhanced corrosion of the Zircaloys is observed, which turned out to be mainly influenced by the given fast neutron flux.

The growth rate of uniform oxide layers in a BWR environment is rather low due to the low system temperatures and can be neglected (Fig. 2). Discussions and experimental programs focus on nodular corrosion, which increases with burnup according to a power law $\sim (BU)^{0.7}$. The axial profile mainly follows the burnup (flux) profile. Sensitively the nodular corrosion depends on the material conditions. Many results point to the size and distribution of second phase intermetallic particles as composed of Zr(Ni, Fe) and Zr(Cr, Fe) as being the controlling parameters of nodular corrosion. It varies from reactor to reactor. Highest nodular corrosion was found on materials not β-quenched during their fabrication, whereas β-quenched materials generally showed an improved behavior /3/. For actual practice it is concluded that an appropriate β-treatment is needed and temperature treatments of the material after its last β-quenching should stay within a range in which the distribution of the alloying elements is not affected. The reasons for the reactor-to-reactor variations are still not well known. In-reactor nodular corrosion does not depend on temperature or may even decrease with increasing temperature. Therefore, there is no effect of the heat flux and the oxide layer itself on the corrosion process. Oxide breakaway, but no defects caused by nodular corrosion (disregarding interaction with crud) have been found. There is still no limit known for the allowable degree of nodular corrosion /4/.

The hydrogen pick-up fraction is low in a BWR, and there is no indication of an increase at higher layer thicknesses. For instance, a H_2 concentration of only 50 ppm was found in a BWR fuel rod at a burnup of 45 GW d/t (U) and an average oxide layer thickness of about 60 μm.

2.2 PRESSURIZED WATER REACTOR (PWR) SYSTEMS

Most PWR operate at
- about 350°C max. cladding surface temperature,
- a hydrogen overpressure (dissolved 2-4 ppm) to reduce the oxygen content in the coolant,
- 155 bar system pressure,
- a basic additive (usually 1-2 ppm LiOH, sometimes NH_4OH) to minimize the corrosion and corrosion transport of primary circuit plant materials,
- 0-1200 ppm H_3BO_3 as additive to control reactivity.

In the primary system of PWRs in which the radiolytic formation of oxygen and oxidizing radicals in the coolant is suppressed by the addition of hydrogen to the coolant, the corrosion of Zircaloy was found to be less enhanced by irradiation. The visual appearance of the in-reactor oxide layers and their microstructure are very similar to the ex-reactor products, irrespective of irradiation and heat flux conditions.

Corrosion behavior appears to be strongly dependent on the temperature at the metal-oxide interface. Under heat flux conditions, this temperature increases as the oxide layer thickens, further increasing the corrosion rate. The temperature rise across the oxide depends on its thermal conductivity, which was found to be between 1,5-2,4 W/m·K for unirradiated oxide and to be reduced by ~30% due to irradiation /5/.

A review of PWR fuel rod corrosion results /6/ comes to the following main conclusions:
- In-pile corrosion is characterized (like out-pile corrosion) by a pre-transition and a post-transition regime with cubic and linear kinetics respectively.
- The post-transition corrosion is irradiation-enhanced at layer thicknesses ≥ 5μm. Compared to the applied basic corrosion law (out-of-pile), the enhancement amounts to a factor close to 4 for stress-relief-annealed Zircaloy-4.
- The effect of the oxide layer on the actual corrosion temperature is of high importance; a thermal conductivity of 1.5 W/mK is taken into account. Nucleate boiling has no extra effect on the corrosion behavior.
- The in-pile corrosion results of a quantity of identical material can be characterized by an average value and a Gaussian distribution.
- The absolute oxide thickness is dependent on burn-up, the power history and the thermal-hydraulic characteristics of the plant.

In respect to the allowable oxide thickness of PWR fuel cladding high power experiments showed a loss of integrity of the oxide layer at a certain thickness resulting in local degradation of its thermal conductivity and local perforation of the cladding wall. Experience with thick oxide layers showed a strong influence of the heat flux (Fig. 3) on the allowable oxide thickness /5/: At a heat flux of 70 W/cm2, Eddy Current (EC) signals were detected at fuel rod positions with an average oxide layer thickness of > 115 μm around the circumference. Perforations and strong EC signals occurred only at positions with > 140 μm oxide thickness on circumferential average, which corresponds to > 160 μm at the local maximum.

In respect to hydrogen up-take it can be concluded /5/6/ that the amount of hydrogen absorbed by the cladding slowly increases superproportionally after having reached oxide layer thickness as of more than 20 μm. Precipitation of zirconium hydrides occurs at hydrogen concentrations above about 200 ppm, starting at the coldest positions. Such positions are the outer rim of the cladding wall, that section of the circumference with lowest oxide thickness, axial gaps in the fuel column, and combinations of these /6/. At an oxide scale thickness of 100 μm the average hydrogen-concentration in the cladding reaches 500 ppm.

3. OXIDATION UNDER ACCIDENT CONDITIONS IN STEAM

Zircaloy-4 oxidation under accident conditions was verified by experiments exposing fuel cladding tubes to steam at temperatures derived from calculated loss-of-coolant (LOCA) transients and those peaking at even higher temperatures and causing severe fuel damage (SFD). The experiments evaluated by means of gravimetry and metallography were able to describe the kinetics of mass increase (mainly by oxidation) and formation of oxygen-containing surface layers (Fig. 4). The simultaneously occuring effects - as hydrogen and heat production, change of cladding tube dimensions and mechanical properties by oxidation, besides enhanced oxidation by creep deformation - equally were determined.

3.1 ISOTHERMAL OXIDATION KINETICS

In respect to PWR loss-of-coolant accidents the isothermal oxidation of Zircaloy-4 cladding material has been investigated in steam within a temperature range of 600-1300°C and an exposure time of ≤ 15 min /7/. The kinetics of oxygen uptake, ZrO_2-scale, and α-Zr(0)-layer growth was expressed by simple rate laws: (below 900°C) of cubic, (above 900°C) of parabolic time and exponential temperature dependences. The equations, deduced from these results describe with good approximation the kinetics of oxygen uptake (ι), the growth of the oxide (ψ), (α), and of the oxide plus α-phase double layer (ξ). High-temperature oxidation tests at 1350-1600°C /8/ have been performed under the aspect of accident conditions leading to severe fuel damage. By a suitably controlled initial exposure to steam the specimen temperature could be stabilized at its desired high level so that an evaluation with respect to oxidation kinetics became feasible.
　　During the extension of the tests from 15 min to 25 h at 600-1600°C (Fig. 5) a special study of the so-called breakaway effect could be performed (Fig. 6). This term stands for the loss of protectiveness of an oxide scale due to its mechanical failure and consequently the transition to faster oxidation, expressed (as mentioned earlier) by a linear rate law. Breakaway oxidation - when steam in surplus was supplied - went along with strong hydrogen pick-up. Within the range of medium test temperatures the initial oxidation kinetics are governed by cubic to parabolic rate laws which change towards linear functions after having reached the critical oxide layer thicknesses. Within the (α + β) - Zr temperature range the effect of the breakaway on oxidation kinetics was found to be moderate. At 1050°C and above parabolic kinetics persist. The metallic matrix under consumption and the growing tetragonal/cubic oxides are sufficiently plastic to accomodate one another (Fig. 6), thus avoiding the build-up of the compressive oxide growth stresses which otherwise induce the breakaway. Considerable changes in tube dimensions have been observed: After total wall oxidation by double-sided steam exposure during 6 h at 1200-1300°C an increase of the outer diameter by 12%, inner diameter by 8%, and wall thickness by 30-60% was measured.

3.2 TEMPERATURE-TRANSIENT OXIDATION KINETICS IN STEAM

Under temperature-transient conditions the oxygen uptake was mainly determined by the duration and the temperature of exposure. Therefore, in all cases of transient temperature exposure, the results have shown that under the chosen reduced time-at-temperature conditions, the extent of oxidation was lower than under isothermal ones. A survey on all results of the applied temperature transients in comparison to our isothermal data and those of Baker-Just (Fig. 7) shows that the measured weight gain, when plotted as function of final plateau temperature /9/, reveals obviously considerable conservatisms in the currently used licensing practice. Various experiments were performed in which - after having reached the critical breakaway oxide scale thickness by isothermal oxidation - slow heating or cooling ramps into/or out of/or by avoiding the breakaway existence range were performed. The existence range of the breakaway effect was found to correspond to the prediction based on the isothermal investigation.

3.3 CREEP-RUPTURE BEHAVIOR IN STEAM

By isothermal-isobaric creep and creep-rupture testing of internally pressurized tube capsules (600-1300°, ≤ 150 bar) /10/ which were externally exposed to steam, the interaction of creep deformation and oxidation was determined. The experiments were interrupted and evaluated when either different states of creep deformation were attained or rupture occurred, or minor or major wall consumption by oxidation had taken place.

One result in respect to the accident behavior of Zircaloy-4 cladding indicates the existing limitation with respect to the creep strength of the one-sided highly oxidized material. By one-sided oxidation of tube capsules (1-360 min, 1000°C, steam) the influence of oxide scale growth on creep strength was elaborated by creep-rupture tests in steam. It could be shown that after having surpassed a temperature dependent critical extent of oxidation, the initial metal/oxide compound-related gain turns over to a total loss of strength and brittle fracture. The critical values-as shown also in Fig. 8- are given in the following Tab.1.

Table 1: Conditions Leading to Instant Brittle Rupture by Creep Testing in Steam

State of Pre-Oxidation			Test Conditions	
Ox. Thickness / [μm]	Metal Wall Consumption [μm]	[%]	Temperature / [°C]	Burst Pressure [bar]
350	227	30	1000	≥ 28
200	130	17	1200	≥ 18

Beyond the above given test conditions the internal pressure build-up leading to instant rupture drops below that of non-preoxidized tube material and finally to zero. In principle, these findings are in agreement with the Emergency Core

Cooling Criteria (ECCC), limiting the maximum cladding surface temperature to 1200°C and the maximum percentage of wall consumption to 17%=125 μm.

4. CONCLUSIONS

The potential limiting parameters for Zircaloy cladding corrosion in conditions of extended burn up of LWR fuel have been considered. Under PWR operation conditions there is no strong argument against a moderately extended burnup in spite of an at least linear increase of corrosion as consequence of increasing surface temperature at the metal-oxide interface. Under BWR operation conditions however nodular corrosion - as a localized corrosion phenomenon - cannot yet be suppressed by means of a suitable material conditioning, making extrapolations to long-time behavior complicated.
Its allowable degree is not yet clearly defined. Additionally, considerations of extended burn-up should take into account - besides waterside corrosion - the extent of cladding oxidation during a hypothetical LOCA which is limited by the ECC-criteria to 17% = 125 μm of wall consumption. This limitation was verified experimentally in respect to brittle fracture when after extensive preoxidation Zircaloy tube capsules were creep-rupture tested in steam.

5. ACKNOWLEDGEMENTS

Substantial help to summarize the experience on Zircaloy water corrosion under LWR operation conditions was given by Mr. F. Garzarolli from Siemens KWU Erlangen and is gratefully acknowledged.

6. LITERATURE

/1/ M.O. Marlowe, J.S. Armijo, B. Cheng, R.B. Adamson, Proc. ANS Intern.Topical Meeting on LWR Fuel Performance, April 21-24, 1985, Orlando, Fla, USA, p. 3/73-90

/2/ F. Garzarolli, D. Jorde, R. Manzel, G.W. Parry, P.G. Smerd (CE/KWU/EPRI), "Review of PWR Fuel Rod Waterside Corrosion Behavior", RP 1250-1 · Taks A · Combustion Engineering Inc., June 1979, CE-NPSD-79; August 1980, EPRI NP-1472

/3/ F. Garzarolli, H. Stehle, E. Steinberg, H. Weidinger, Proc. ASTM - 7th Intern. Symp. on Zirconium in the Nuclear Industry, June 24-27, 1985 Strasbourg (France), to be published

/4/ F. Garzarolli, R. Manzel, Proc. KTG-Reaktortagung Mannheim, Germany 1977, p. 477-480

/5/ F. Garzarolli, H. Stehle, Proc. IAEA-Symp. on Improvements in Water Reactor Fuel Technology and Utilization, September 15-19, 1986 Stockholm (Sweden), p. 387-407

/6/ F. Garzarolli, R.P. Bodmer, H. Stehle, S. Trapp-Pritsching, Proc. ANS Meeting on LWR Fuel Performance, Orlando, Fla, USA Vol. I (1985) 3/55-72

/7/ S. Leistikow, G. Schanz, H. v. Berg, KfK-Rep. No. 2587 (1978)

/8/ S. Leistikow, G. Schanz, H. v. Berg, A.E. Aly, Proc. OECD-NEA-CSNI/IAEA Meeting on LWR Fuel Safety and Fission Product Release in Off-Normal and Accident Conditions, May 16-20, 1983 Risø (Denmark) 188-199

/9/ S. Leistikow, G. Schanz, Werkst. Korr. 36 (1985) 105-116, Nucl. Engg. Des. 103 (1987) 65-84

/10/ S. Leistikow et al., KfK-Rep. No. 2750 (1979) 4200/48-70

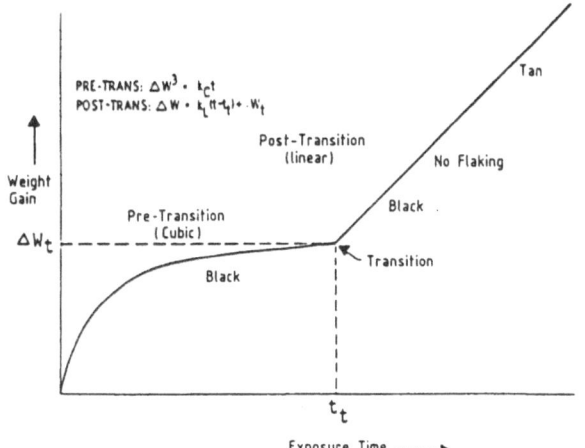

PRE-TRANS: $\Delta W^3 = k_c t$
POST-TRANS: $\Delta W = k_L (t-t_t) + W_t$

Post-Transition
(linear)

Tan

No Flaking

Weight
Gain

Pre-Transition
(Cubic)

ΔW_t

Black

Black

Transition

Black

t_t

Exposure Time ⟶

1   Pre- and Post-Transition Periods during
     Zircaloy Oxidation in Water or Steam
     (250-400°C)

Oxide Thickness (average lift-off) at peak position

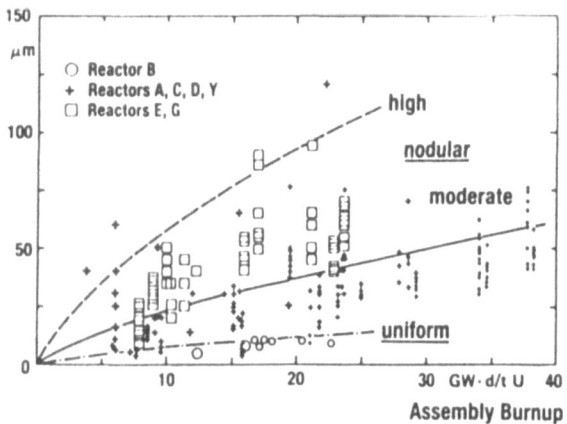

150
μm

○ Reactor B
+ Reactors A, C, D, Y
□ Reactors E, G

high

nodular

100

moderate

50

uniform

10   20   30   GW·d/t U   40

Assembly Burnup

2   Corrosion of Zircaloy-2 BWR Fuel Rods,
     Produced 1965-1980 (after Garzarolli and
     Stehle /5/)

Oxide Thickness

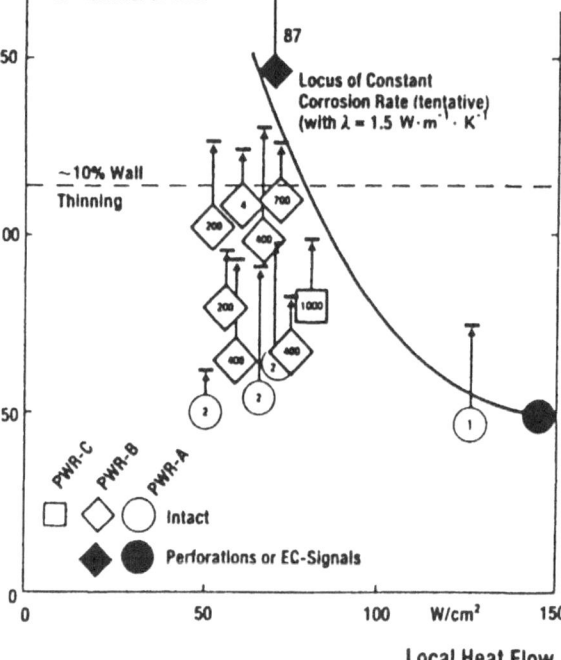

200
μm

⊥ Circumferential max.
Ⓧ Average
x = Number of Rods

150

87

Locus of Constant
Corrosion Rate (tentative)
(with $\lambda = 1.5$ W·m⁻¹·K⁻¹)

~10% Wall
Thinning

100

50

PWR-C PWR-B PWR-A

□   ◇   ○   Intact

◆   ●   Perforations or EC-Signals

0
0   50   100   W/cm²   150

Local Heat Flow

3   Map of Experience with Thick ZrO_2 Layers in
     PWRs. Oxide Thickness Versus Local Heat Flux
     (after Garzarolli and Stehle /5/)

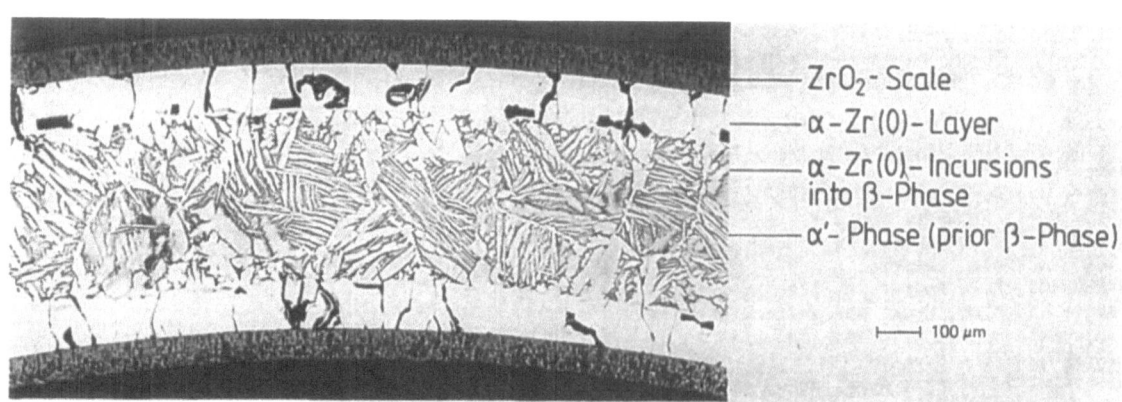

ZrO_2- Scale
α-Zr(0)-Layer
α-Zr(0)-Incursions
into β-Phase
α'- Phase (prior β-Phase)

⊢—⊣ 100 μm

4   Cross-Section of Zircaloy-4 Tubing when
     Double-Sided Oxidized in Steam (2 min,
     1400°C)

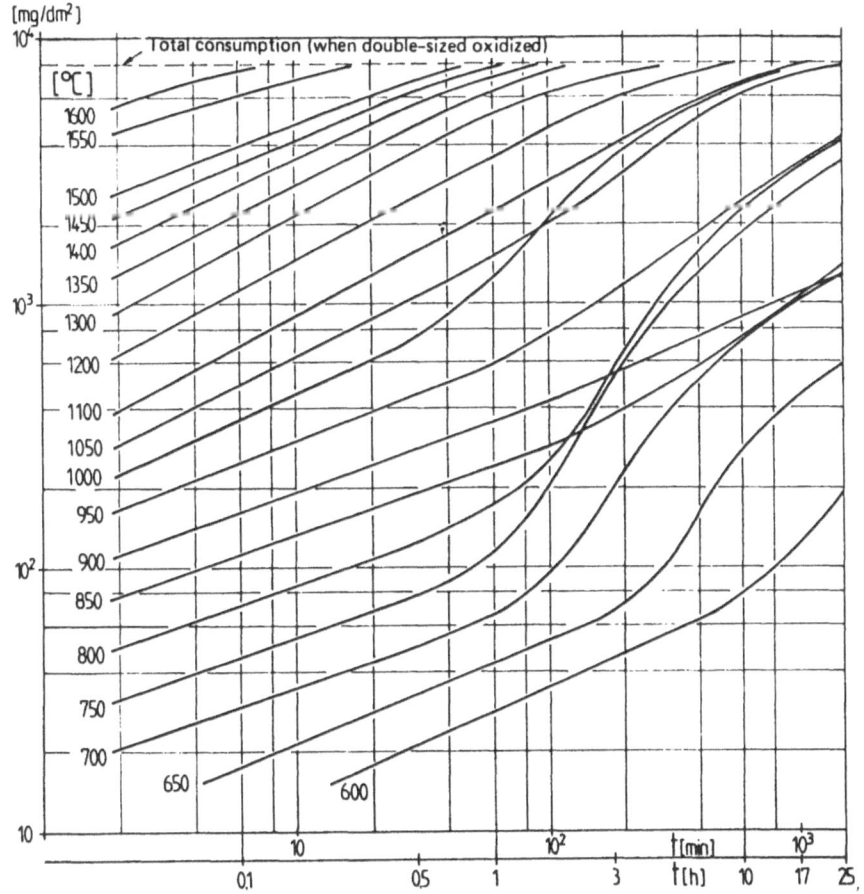

5   Mass Increase of Zircaloy-4 Tubing versus
     Time of Exposure in Steam (≤ 25 h, 600-
     1600°C)

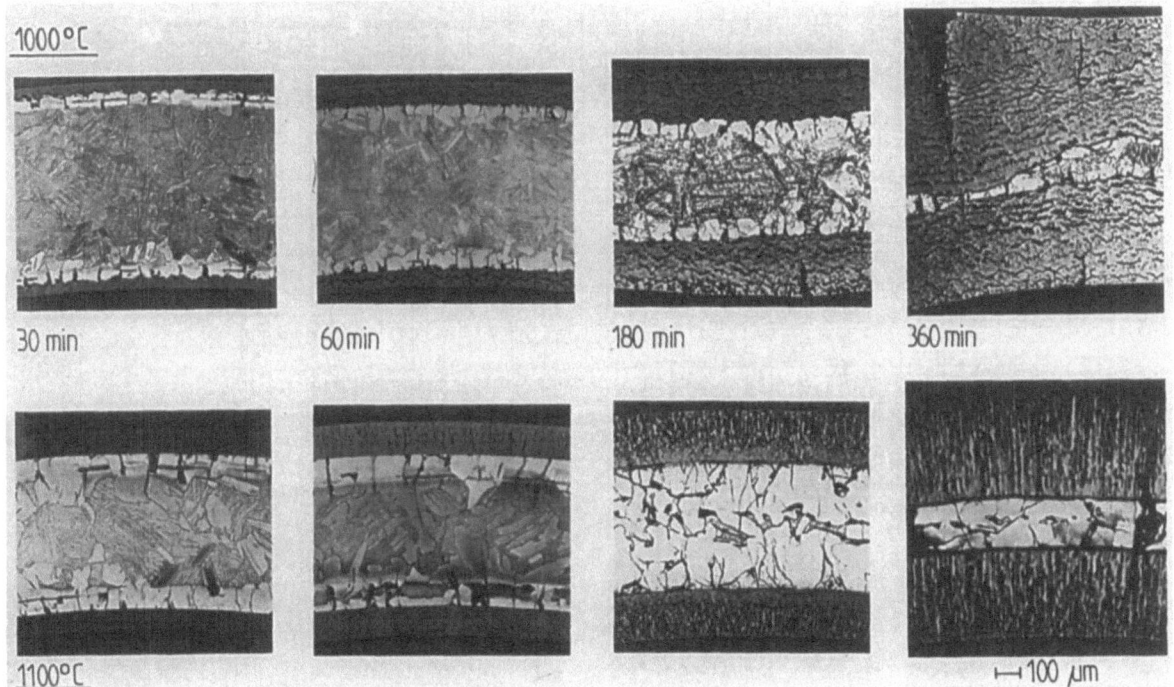

6   Growth of Breakaway and Adherent ZrO_2
     Layers in Steam (1000 and 1100°C, 30-360
     min)

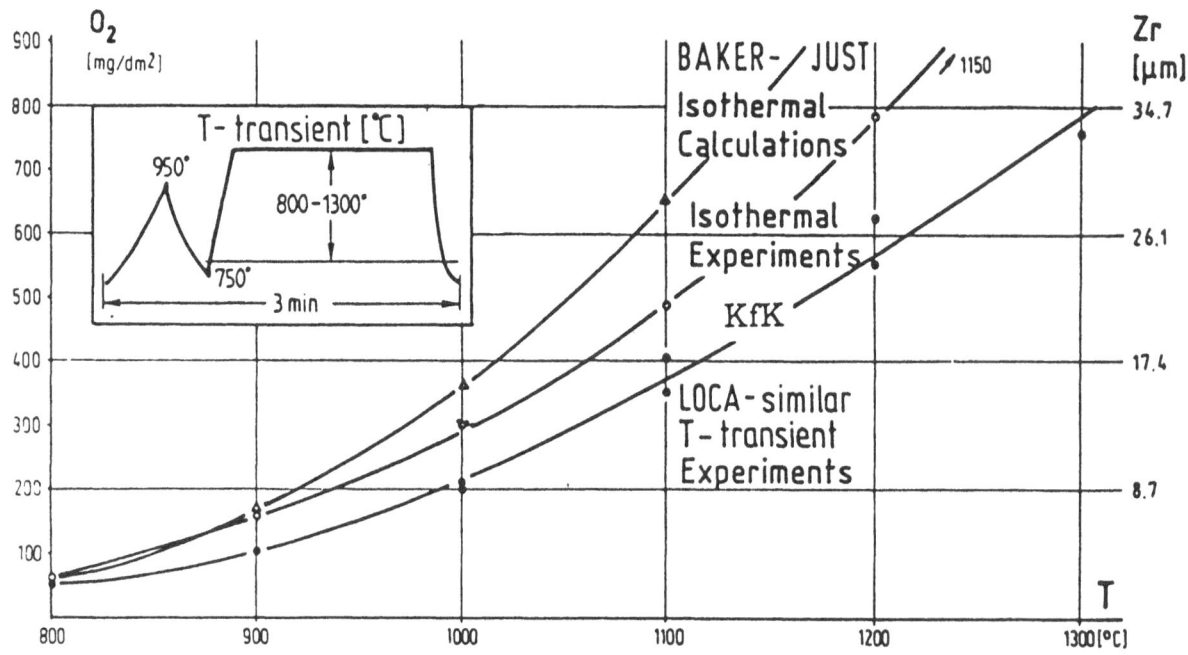

7   Comparison of Mass Increase of Zircaloy-4,
as Calculated According to Baker-Just, as
Measured by Isothermal and LOCA-Transient
Exposures in Steam (3 min, 800-1300°C)

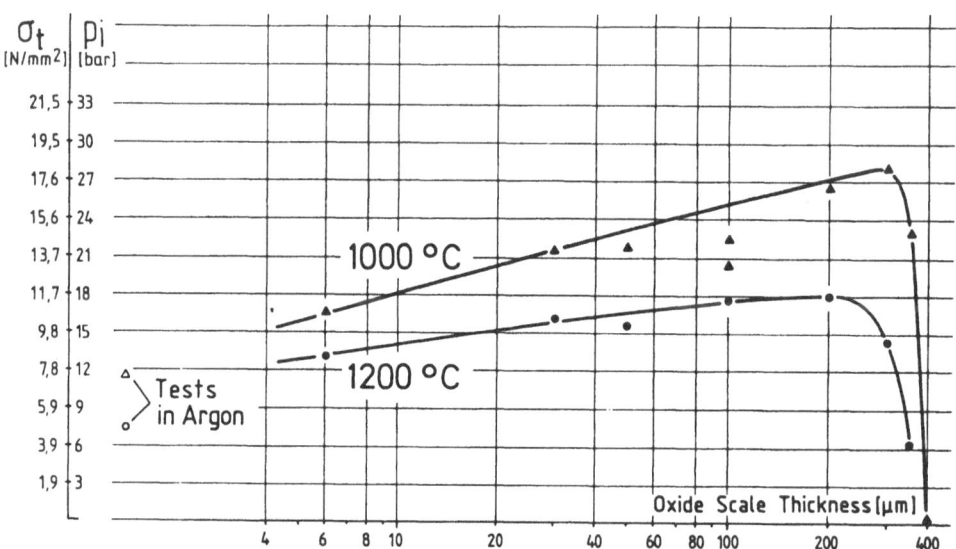

8   Creep-Rupture Experiments of Zircaloy-4
Tube Capsules in Steam at 1000 and 1200°C
after External Preoxidation (1-360 min,
1000°C, Steam)

4 The Oxidation of Structured Stainless Steels and Fuel Cladding in the Advanced Gas-Cooled Reactors

H. E. Evans

Berkeley Nuclear Laboratories,
Central Electricity Generating Board,
Berkeley, Gloucestershire, GL13 9PB.

SYNOPSIS

This paper surveys the oxidation behaviour of the whole range of stainless steels used in the Advanced Gas-cooled Reactors (AGRs). The steels are categorised into: 18Cr varieties containing between 9 and 12Ni which are used for structural items and boiler tubes; 25Cr/20Ni steel used as thermal insulation foils; 20Cr/25Ni alloys used as fuel element cladding. The maximum operating temperatures vary from ~923K for the structural steels and insulation foils to ~1143K for fuel cladding.

An important feature which determines performance is the metal section lost due to oxidation. To assess this, proper allowance must be made for localised as well as uniform attack and, as a result, much emphasis is placed on the ability of the steel to form healing oxide layers. Under normal operating conditions, no restrictions are anticipated as a result of oxidation. A detailed discussion is provided of the principal postulated faults within the AGR system. The extent of clad oxidation under such circumstances is an important factor in assessing fuel pin integrity.

INTRODUCTION

Stainless steels have an important structural role within the circuit of the Advanced Gas-cooled Reactors (AGRs) and find duty as fuel element cladding, insulation foils, within the boilers and in other locations. The diverse alloys used vary considerably in microstructure, composition and, in some important aspects, in oxidation behaviour. The most onerous exposure conditions exist for fuel cladding and particular care must be taken in assessing its behaviour during both normal operation and potential fault conditions.

COMPOSITION AND PRINCIPAL USES

The compositions of the various stainless steels employed in the AGRs are outlined in Table 1. Some of the principal components made from these, together with representative service conditions are given in Table 2. In all cases the steels are exposed to the coolant gas $(CO_2/1-1.5^V/oCO$ containing smaller quantities of CH_4, H_2 and H_2O) at 4.1 MNm^{-2} pressure, apart from Dungeness 'B' which operates at ~3MNm^{-2} pressure.

Dwell periods and operating temperatures vary considerably. The shortest dwell is seen by the fuel cladding at a maximum of 40,000 hours but this component also experiences the highest operating temperature of ~1143K. On the other hand, structural steels must remain in situ for the reactor lifetime (~250,000 hours) but experience correspondingly lower operational temperatures (typically ~923K maximum). In general, the more highly alloyed steels are used where temperatures are high and/or dwell periods long or where component redundancy is limited.

The steels shown in Table 1 will be categorised into 18Cr structural steels, 25Cr insulation foils and boiler support items and the 20Cr fuel cladding alloys. Their oxidation behaviour will be considered in this order.

15

OXIDATION CHARACTERISTICS OF THE 18Cr STRUCTURAL STEELS

General Kinetics

The 18Cr steels given in Table 1 are here described as a group with characteristics which can be applied to both the wrought and cast forms. Previous reviews of the oxidation behaviour of these steels for AGR application have been provided in references (1-4). The range of kinetics exhibited by these steels at their peak operating temperature of 923K is depicted in Figure 1 and is divided into three broad types.

Type 1 kinetics are portrayed by those components, mainly machined or cast, which have excellent oxidation resistance. The oxide formed is thin, uniformly distributed and chromium rich. Type 2 kinetics are followed by the majority of reactor steels. They are characterised by a rapid decrease of a high initial rate of attack to a final value comparable with those found in Type 1 kinetics. The oxide formed during these early stages is duplex in nature (Figure 2), composed of an outer layer of magnetite and an inner iron-chromium spinel containing $\sim$25^W/o chromium. The decrease in rate at longer times is associated with the formation of a chromium- and silicon-rich healing layer at the base of the spinel (Fig. 2). Type 3 kinetics are an accentuated form of Type 2, in that a healing layer eventually reduces the rate of duplex oxide penetration. The extended period of duplex oxidation in this case is achieved, however, by the introduction of an excessively large metal grain size or by artificially depleting the surface in chromium. Such Type 3 kinetics are not characteristic of production reactor material.

It can be appreciated from this discussion that the feature that differentiates between kinetic type is the stage or depth into the steel at which a healing layer is formed. For a given intrinsic rate of duplex oxidation, the ease of healing is determined by the rate at which chromium and silicon ions are supplied from the metal to the oxide/metal interface. This transport is aided by cold work (diffusion short circuiting by dislocations) and by a fine grain size (grain boundary short circuiting). It is for this reason that cold worked steels, which includes all cast components, and those with fine grains tend to display Type 1 kinetics or lie towards the lower bound of Type 2. For those examples which do not form a chromium-rich layer at the surface, duplex oxidation at 923K tends to be stopped at the first internal grain boundary (e.g. Figure 2) due to rapid supply of chromium and silicon to this interface.

At temperatures appreciably less than 923K (e.g. 800K) the rate of transport within the steel is always insufficient to form healing layers even at grain boundaries. Because of this, the metal loss rate increases with decreasing temperature to reach a maximum at around 790 to 830K. This effect for relatively short exposure periods is shown in Figure 3(4).

As a final comment, it should be noted that increasing silicon content of the alloy within its specification range has a beneficial effect on the oxidation resistance (1). On the other hand, removal of a protective oxide by thermal spallation or mechanical abrasion can trigger duplex oxidation of the chromium-depleted substrate in that region. Under duplex oxidation conditions, in general, carbon becomes injected into the steel from the CO_2-based oxidant and precipitates as chromium-rich carbides. However, provided that the component section thickness is large (as will be the case) negligible overall depletion of chromium within the bulk steel occurs, and the ability to heal this form of attack is not impaired. By contrast, contaminants such as chlorine and chloride ions can have a major deleterious influence. These appear to impair the ease with which healing layers can form and encourage widespread development of duplex oxidation. A result is that enhanced oxidation rates can continue for many thousands of hours exposure. Because of this, much care is taken to exclude chlorine-containing compounds from the reactor circuit.

Predictions of Behaviour in Reactor

In order to demonstrate the safe and reliable operation of high-temperature structural components, it is necessary to calculate the expected loss of metal section at end of service life (250,000 hours). To do this, the weight gain kinetics, such as Type 1 and 2 of Figure 1, are expressed in a fractional power law of time to aid extrapolation. From the weight gain, a uniform oxide thickness is calculated and a maximum value of metal section loss by pitting estimated by multiplying this by a factor in the range 1.3-2.4. The mean metal loss values calculated at end of life (250,000h) at various exposure temperatures are shown in Figure 4 (ref. 4) for steels of differing grain sizes. At those temperatures, e.g. 923,973K, at which healing layers readily form, the metal loss predicted has a value similar to the alloy grain size. This is consistent with the metallographic evidence described above (e.g. Figure 2) which shows that healing layers can form at interior alloy grain boundaries. At lower temperatures, the expected section loss increases as it becomes more difficult to heal the duplex oxidation. The maximum rate of section loss is dependent on alloy grain size but occurs at an intermediate temperature (e.g. at 873K for >80 μm grain size but at $\sim$793K for finer grains) within the operating range. At temperatures of 793K and less there is no effect of grain size.

The values of section loss shown in Figure 4 are, generally, very much less than the component section and no potential problems or restrictions have been identified. The one exception is the case of insulation foils (which are of relatively thin section) for which the use of 18Cr steels has been restricted to temperatures less than 673K. Confidence in the validity of these predictions is maintained by oxidation testing, which currently extends to 100,000h, i.e. well in advance of station operation. In addition, monitoring of the behaviour of bolted, welded and dowelled joints is undertaken to ensure that no anomalous behaviour arises in such interfacial situations. These and coupon samples are located both in the reactor circuit and in on-site autoclaves (5).

OXIDATION CHARACTERISTICS OF 25Cr STEELS

These steels, specifically as AISI Type 310, are used as thermal insulation foils for the reactor vessel in locations where they will experience temperatures >673K and also as boiler structural items. The thinnest foil sections are $\sim$100 μm. A review of the oxidation properties of these steels with reference to this AGR application is given in ref. (6).

The oxidation behaviour at the peak operating temperatures of 873-923K can be represented by the Type 1 and Type 2 kinetics of Figure 1 provided that the right-hand ordinate axis is used. It can then be appreciated that the rate of reaction is up to a factor 10 less than the equivalent rates in the 18Cr steels. A greater propensity exists for Type 2 kinetics to be followed at 923K than at 873K. The highly protective Type 1 behaviour is associated with the formation of a chromia layer, whereas Type 2 reflects duplex oxidation but not to the same extent as in the 18Cr steels. Nevertheless, duplex oxidation is again associated with coarser-grained materials.

Using similar extrapolative techniques to those used in the 18Cr steels, the predicted mean section loss at 250,000h due to the uniform surface oxide is only 3.5 μm at 923K and 1.0 μm at 873K. However, the principal loss of section found in ref. (6) was due to intergranular oxidation below the uniform oxide. Such attack was found under all but the most severe form of duplex oxidation. These penetrations were found to be chromium rich but their rate of growth decreased very rapidly with time. Work subsequent to that reported in (6), (M.G. Angell, private communication) suggested that these intrusions represented rolling defects which had subsequently oxidised but did not propagate. Even so, assuming a non-zero growth rate of these penetrations gave a predicted mean section loss after 250,000h at 923K of ~15 μm. This value also corresponds to the section lost during severe duplex oxidation for which the intergranular attack was absent. In either case, the predicted losses are appreciably less than the thickness of even the thinnest foils used.

20Cr FUEL CLADDING

Oxidation under Normal Operating Conditions

Fuel cladding in the AGRs experience operating temperatures ranging from 673 to 1143K depending on axial and radial location within the reactor core. The maximum temperature of 1143K exists on only a small fraction of pins (< 1%) and then only in a localised region on the pin. Even so, such high temperature regions will be sustained for a significant fraction of fuel life and will decrease mainly with fuel reactivity. A description of the oxidation behaviour of 20/25/Nb cladding with emphasis on its spallation behaviour has been given elsewhere (7). The present paper concentrates on those aspects which relate directly to fuel endurance under fault conditions.

The oxidation characteristics of fuel cladding tend to be of practical significance to pin endurance only for operating temperatures >1023K and it is this domain of temperature which will be considered here. Most of the emphasis will be on the current standard Nb-stabilised 20Cr/25Ni steel (Table 1) although a TiN-dispersion strengthened alternative alloy is available for the system should increased operational flexibility, such as load following, be required.

In common with the other stainless steels, the ease of formation of the initial protective chromia layer depends on the level of cold work and surface grain size. Since the clad steel is used in the annealed state, surface grain size becomes the pre-eminent factor. This grain size is relatively small (8 μm) so that, at the temperatures of interest, a protective chromia layer forms very rapidly: in ~10h at 1023K and ~1h at 1123K (7). Prior to loading into reactor, any UO_2 contamination on the fuelled pin is removed by electrolytic cleaning in an aqueous nitric acid solution. This treatment also removes the deleterious effects of surface roughness which may have been introduced during can machining (8).

The subsequent oxidation of these steels occurs under a substantially protective oxide which thickens according to parabolic kinetics, e.g. Figure 5 (9). The rate of reaction is insensitive to the pressure of the oxidant in the range 0.1 to 4Mnm⁻² , also to CO content in the range 1 to 2^v/o and to small additions of methane and water vapour (7). Whilst the oxide formed is chromium-rich, it also contains appreciable amounts of manganese (as a spinel), particularly near the gas interface, and is separated from the metal by a thin (~10nm) layer of amorphous silica (10,11). With increasing exposure, especially at 1123-1173K, the iron content of the film increases and the weight gain kinetics depart from parabolic. Even so, the metal section loss due to the formation of the uniform oxide is relatively small even at these high temperatures. Thus, an exposure of 40,000h at 1143K will remove only ~11 μm of metal (Table 3) and this loss poses no significant threat to the integrity of the clad of 380 μm wall thickness. The application of these out-of-reactor results to the high-flux region of the reactor core is supported by experimental work which shows negligible effect of neutron and fission fragment irradiation on oxidation rates at temperatures of current interest, e.g. 1073K (13). The protective conditions described above result from the selective oxidation of chromium and silicon to form surface layers. These are the elements which have been identified earlier as being of similar importance to the structural stainless steels. However, a consequence is that the metal in the vicinity of the surface oxide becomes depleted in these elements and, ironically, thereby increases the susceptibility of the steel to severe localised oxidation.

Such local attack occurs when the surface oxides fail mechanically, either through cracking or spalling. A result is that the chromium-depleted metal substrate becomes exposed to the oxidant and rapidly develops iron-rich oxides in the form of a localised pit (10). The pit penetrates into the steel at a rate many orders of magnitude faster than that of the original protective oxide and constitutes a potent mechanism for consuming metal. Fortunately, as the pit penetrates into the steel, the local chromium concentration at its base increases, simply because the extent of chromium depletion under the original protective oxide decreases with depth into the metal. Eventually the pitting reaction reaches a depth at which the local chromium concentration is sufficient to re-form a protective chromia layer. At this stage, the pit propagation rate drops markedly and subsequent loss of section is negligible. An example of healed pits in the standard cladding alloy is shown in Figure 6. The larger of the two pits shown extends ~40 μm into the steel and is some 150 μm in breadth. These dimensions are considerably greater than the metal grain size (8 μm) and demonstrate that this form of high-temperature pitting is not healed easily at internal metal grain boundaries as was the case for 18Cr steels at lower temperatures. This difference arises because the rate of propagation of the pit is high (~1 μm/hour) and is too high even for grain boundary transport of chromium to be effective over long distances.

Figure 6 also serves to show that a wide distribution of pit depths can exist on any particular specimen. This arises principally because pits are formed at different times during the exposure and propagate to depths characteristic of the chromium depletion profile at that time (12). Measurements on healed pits indicate that healing occurs at a depth into the depletion profile corresponding to ~16^w/oCr. This concentration contour penetrates the steel approximately parabolically with time and so shows that the maximum depth of pitting will also increase parabolically. The temperature variation of this maximum depth will reflect the temperature dependence of the chromium-depletion profile. Best estimates of this maximum depth of pitting attack are given in Table 3 for an exposure of 40,000h at various temperatures. Also shown is the loss of section expected from protective uniform oxidation only.

The important feature of this comparison is that even though pitting attack results in a significant loss of section, the maximum depths predicted still constitute only 12% of the clad section even at the highest operating temperature. Extended exposures at temperatures higher than this, e.g. 1173K will lead to general loss of protectivity of the surface oxide (NB. the incorporation of iron into the oxide layer as mentioned above). A consequence is the development of internal oxidation in the forms of silica intrusions with associated carburisation. This is the temperature range in which the steel begins to suffer non-selective oxidation and within which its oxidation resistance rapidly declines. The current aproach is to support the operation of fuel cladding by advance laboratory testing in order to ensure that such a breakdown regime is not entered.

The rate of uniform oxidation of the TiN-strengthened, alternative cladding alloy is similar to that found on the Nb-stabilised steel (7), as shown in Figure 5. However, the alloy experiences a greater extent of localised oxidation and, overall, exhibits poorer oxidation properties than the standard cladding alloy. Nevertheless, oxidation under normal operation poses no threat to cladding integrity.

This description has emphasised that both the uniform and local loss of section experienced by fuel cladding is insufficient to promote pin failure by oxidation during normal operation. As will be seen, however, the levels of attack predicted are still significant under fault conditions and their effect must be taken into account in safety analyses.

Behaviour under Fault Conditions

Types of Fault

Postulated faults in the AGR system are categorised into: (a) depressurised faults, which involve either a complete or partial loss of coolant; (b) pressurised faults, which take place whilst the reactor coolant is at full pressure. Numerous different faults exist within each category but the most severe of each will be considered in this section. The details apply specifically to the Hinkley Point 'B' reactors but have relevance to other AGRs also.

Clad Behaviour during a Depressurisation Fault

The most severe depressurisation fault is envisaged to occur as a result of a major breach in the CO_2 bypass circuit (e.g. ref. 14). As a consequence the gas coolant is lost at its fastest rate and the reactor is totally depressurised in 30 minutes from the onset of the fault. Fuller details of the post-fault sequences can be obtained from ref. (14). For present purposes, the important point to note is that significant cooling ability is maintained during the depressurisation sequence and, as a consequence, the clad temperature increases by only 60 deg.K from that which existed at the start of the fault.

A result of the relatively low clad temperatures experienced in the AGR depressurisation is that negligible oxidation occurs during the fault. The importance of the extent of clad oxidation prior to the fault then lies in its effect on tube rupture behaviour of pins during the fault. It is when the reactor is fully depressurised and clad temperatures are relatively high that pins accumulate most creep damage. This arises simply because the internal gas pressure within the pin exerts a nett tensile stress in the clad wall which is at a maximum when the circuit pressure is lowest. Prior oxidation is assumed to reduce the load-bearing section of the clad uniformly by an amount equal to the maximum pit depth predicted, e.g. Table 3. For the 20/25/Nb steel (cf. Table 1), the pin rupture life varies as (wall thickness)^5, so that, for a given pin internal pressure, a 50 μm pit in the original 380 μm wall thickness is calculated to reduce rupture life by a factor 2. Thus, even the relatively modest levels of attack predicted prior to the fault (Table 3) can have a significant influence on behaviour during the fault. Unlike structural steels, it is not sufficient to conclude simply that satisfactory behaviour will obtain if the fraction of section lost under normal operation is small. The behaviour of pins during the fault imposes a much more severe acceptance criterion and is a principal consideration when designing reactor output and fuel dwell period.

One advantage of the TiN-strengthened alternative cladding is that its superior creep properties result in far greater margins in this fault, even after allowing for its inferior oxidation behaviour during normal operation, than exist with the standard steel.

Clad Behaviour during a Pressurised Fault

The most severe pressurised fault envisaged is the unintentional withdrawal of control rods from the reactor core leading to rapid local rises in reactivity and clad temperature. Details of such faults and the protection systems used in the AGR are given in ref. (14). It is emphasised in that paper that a conservative absolute limit on maximum clad temperature permitted during the fault, for even short periods, is 1623K (1350°C) in order that clad melting does not occur. In addition, the high temperatures involved could produce significant clad oxidation during the fault, in contrast to the depressurisation fault discussed above. It is necessary to demonstrate that not only will clad not melt but that its section lost due to oxidation will not prejudice the ability to maintain core cooling nor to remove fuel from the core when the fault situation has been terminated.

18

Studies of the oxidation properties of AGR cladding are currently being made in laboratory to temperatures of 1623K (1350°C) and initial results have recently been reported (15). The main observation at these temperatures is that protective oxidation by the formation of chromium- and silicon-rich layers is difficult to maintain and that the oxidation reaction quickly degenerates into a non-selective form of attack in which iron is also oxidised. An example of a section of clad tested in the laboratory and held at 1473K (1200°C) for 1 hour in notionally $CO_2/2^V/oCO$ is shown in Figure 7. The oxide morphology is essentially the same as that found at this temperature by Leistikow(16) for No.1.4970 stainless steel (15Cr/15Ni) tested in steam and by Ishida et al.(17) for Type 304 steel, again tested in steam.

The experimentally observed kinetics of oxidation obtained isothermally are used by integrating numerically through the temperature/time transient of the fault. Clad section loss is then calculated as a first stage in assessing the mechanical endurance of the fuel both during and post fault.

As a demonstration of the ability of the clad to maintain integrity during such transients, a severe pressurised fault was simulated in a series of four experiments in the Windscale AGR using both new and previously irradiated fuel equipped with thermocouples. A typical transient used is shown in Figure 8 (18) and closely resembles the worst predicted for the commercial reactors (14). After each of the experiments, fuel mechanical integrity was preserved and the stringers were discharged routinely. Results of the examination of the extent of clad oxidation are given by Newbigging et al.(19). They show that total depth affected varied from 14μm for experiments 1 and 3 which reached 1553K (1280°C) to 45μm for experiment 2 which reached 1603K (1330°C). There was no apparent systematic difference between new and irradiated cladding and the extent of oxidation is broadly as expected from the recent laboratory tests (R.C. Lobb, private communication).

Laboratory tests have also been undertaken on the TiN-strengthened alloy at temperatures relevant to the pressurised fault. Results are presently limited but indicate a significant reduction in the rate of attack compared with the standard clad for temperatures in excess of ~1400K (15). Such behaviour is consistent with observations (e.g. 20) on other dispersion-strengthened alloys.

An important conclusion to be drawn from this body of work is that the severity of a pressurised fault transient depends on the integration of time at temperature. Thus, short durations at very high temperatures, approaching clad melt, will not prejudice pin integrity for either cladding alloy whereas longer times at lower temperatures could be more damaging and result in greater section loss due to oxidation.

CONCLUSIONS

The stainless steels used in the AGRs have been categorised in this paper into 18Cr structural steels, 25Cr insulation foils and 20Cr fuel cladding. The oxidation characteristics of each group of steels have been considered in relation to safety-related issues. An essential feature in all cases is the calculation of metal section lost as a result of oxidation.

Maximum operating temperatures for both the 18Cr and 25Cr steels are around 923K (650°C). At such temperatures their oxidation behaviour is dominated by the ease with which chromium and silicon-rich protective layers can form. Cold-worked or fine-grained structures readily display protective kinetics with little section loss. On the other hand, annealed, large-grained structures, e.g. near welds, experience relatively non-protective duplex oxidation until a chromium-rich healing layer forms at the interface between oxide and metal. At temperatures of 923 and 873K, this generally occurs at the first metal grain boundary and, as a consequence, end of life section losses are predicted to be little more than the metal grain size. Because of difficulties in forming healing layers at lower temperatures, the extent of attack maximises at some intermediate temperature. Nevertheless, provided that contaminants, particularly halogens, are avoided, no operational constraints are anticipated.

The temperature range of operation of fuel cladding is much wider than for the structural steels and extends to 1143K (870°C). For purposes of endurance assessments it is temperature in the range >1023K (750°C) that is of most significance. Within this range protective oxidation conditions develop within a few hours with little subsequent uniform loss of section. However, mechanical damage or spallation of the surface oxide results in localised oxidation and pit formation. It is this local form of attack, which can penetrate many grain diameters into the steel, which is of significance in a depressurisation (loss of coolant) fault. Assessment must be made in fuel design of the effect of such pits on fuel-pin rupture life during the fault. The AGR core may also experience pressurised faults where reactivity insertions can cause rapid increases in clad temperatures. The extent of clad oxidation under such circumstances becomes an important factor in assessing fuel pin integrity.

ACKNOWLEDGEMENT

This paper was prepared at Berkeley Nuclear Laboratories and is published with permission of the Central Electricity Generating Board.

REFERENCES

1. S.J. Allan, J.F. Norton and L.A. Popple, 'Corrosion of Steels in CO_2', Br. Nucl. En. Soc., London, 1974, p284.

2. J.C.P. Garrett, S.K. Lister, P.J. Nolan and J.T. Crook, 'Corrosion of Steels in CO_2', Br. Nuc. En. Soc., London, 1974, p.298.

3. J.C.P. Garrett, M.G. Angell and A. Whittaker, 'Gas-cooled Reactors Today', Br. Nuc. En. Soc. London, 1982, Vol.2, p.231

4. P.C. Rowlands, J.C.P. Garrett, L.A. Popple, A. Whittaker and A. Hoaksey, Nucl. Energy, 25, (1986), 267.

5. J. Billingsley and A.L. James, Nucl. Energy, 25, (1986), 277.

6. J.C.P. Garrett, J.T. Crook, S.K. Lister, P.J. Nolan and J.A. Twelves, Corros. Sci., 22, (1982), 37.

7. P.W.G. Simpson and H.E. Evans, 'Nuclear Fuel
    Performance', Br. Nuc. En. Soc., London,
    1985, p.265.

8. K.C. Triparthi and J.E. Antill, Corros. Sci.,
    10, (1970), 273.

9. H.E. Evans, D.A. Hilton, R.A. Holm and
    S.J. Webster, Oxid. Metals, 12, (1978), 473.

10. M.J. Bennett, J.A. Desport and P.A. Labun,
    Oxid. Metals, 22, (1984), 291.

11. H.E. Evans, D.A. Hilton, R.A. Holm and
    S.J. Webster, Oxid. Metals, 14, (1980), 235.

12. R.C. Lobb and H.E. Evans, 'Plant Corrosion',
    Ellis Horwood Ltd., Chichester, U.K., 1987,
    Eds. J.E. Strutt and J.R. Nicholls,
    Chapter 15, p.276.

13. M.J. Bennett, G.H. Chaffey and J.E. Antill,
    Corros. Sci., 10, (1970), 273.

14. D.K. Cooper and A.W.A. Gluck, 'Gas-cooled
    Reactors Today', Br. Nucl. En. Soc., London,
    1982, Vol.4, p.197.

15. R.C. Lobb and H.E. Evans, 'International
    Conference on Materials for Nuclear Reactor
    Core Applications', Br. Nucl. En. Soc.,
    London, 1987, p.335.

16. S. Leistikow, 'Zirconium in the Nuclear
    Industry: Sixth International Symposium',
    ASTM STP 824, Eds. D.G. Franklin and
    R.B. Adamson, 1984, p.763.

17. T. Ishida, Y. Harayama and S. Yagachi,
    J. Nucl. Mater., 140, (1986), 74.

18. M.J. Bridge and M. Kendal, 'Gas-cooled
    Reactors Today', Br. Nucl. En. Soc., London,
    1982, Vol. 3, p.81.

19. A.W.P. Newbigging, A.I. Russell and
    E. Turner, 'Fifth International Meeting on
    Thermal Nuclear Reactor Safety',
    Karlsruhe, 1984.

20. C.S. Giggins and F.S. Pettit, Metall. Trans.,
    2, (1971), 1071.

Table 1

Typical Compositions of Stainless Steels

Alloy	Composition, w/o (Balance Fe)							
	Cr	Ni	Si	Mn	C	Ti	Nb	Others
AISI 304	18	9	1.0	2.0	.03	–	–	–
316	18	12	1.0	2.0	.08	–	–	2.0 Mo
321	18	9	1.0	2.0	.08	.40	–	–
347	18	10	1.0	2.0	.08	–	.80	–
AISI 310	25	20	1.5	2.0	.08	–	–	–
20/25/Nb	20	25	0.6	0.6	.05	–	.6	–
20/25/TiN *	20	25	1.0	0.6	.015	2.0	.2	–

*
Subsequently nitrided to produce a through-
thickness dispersion of titanium nitride
particles

Table 2: Typical Operating Conditions and Uses

Alloy Type	Operating Conditions		Principal Components
	Max.Temp. K(°C)	Duration hrs.	
18Cr Structural Steels	923(650)	250,000	Boiler superheaters and reheaters; plug units.
25Cr Steels	923(650)	250,000	Pressure vessel insulation foils; boiler support items
20Cr Fuel Cladding	1143(870)	40,000	Fuel cladding; fuel grids and braces

Table 3

Predicted Depths of Pitting Attack in 20Cr/25Ni/Nb Steel Fuel Cladding after Exposures of 40,000 Hours

Exposure Temperature		Section Lost due to Uniform Oxidation μm	Best Estimate of Max. Pit Depth (Including Uniform Oxidation) μm
K	(°C)		
1023	(750)	3	23
1073	(800)	5	31
1098	(825)	6	35
1123	(850)	9	40
1143	(870)	11	44

21

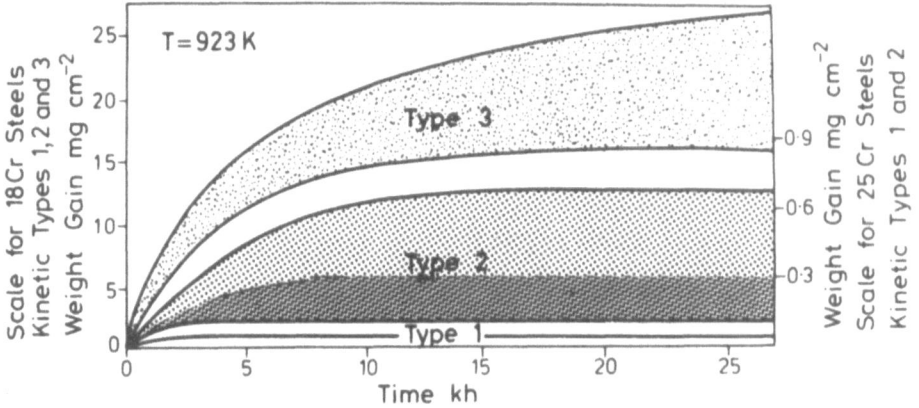

Figure 1: Schematic curves showing Types 1,2 and 3 kinetics for 18Cr steels and Types 1 and 2 for 25Cr steels (derived from refs. 3 and 6.)

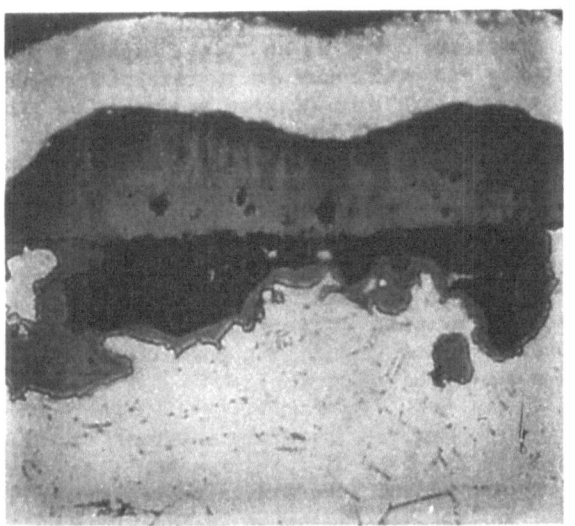

Figure 2: The formation of a healing layer beneath duplex oxide formed on 18Cr steels at 923K (3).

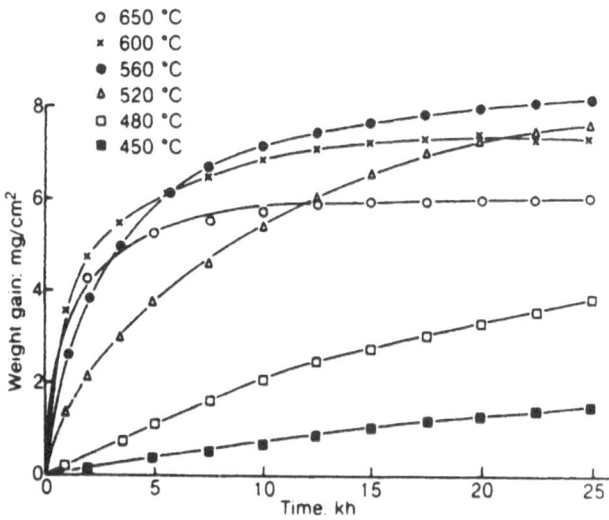

Figure 3: Weight gain kinetics for the oxidation of 18Cr steels at various temperatures (4).

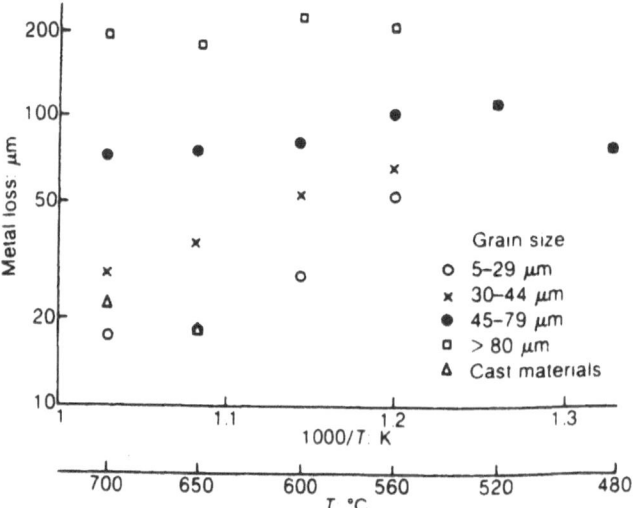

Figure 4: Extrapolated mean metal loss after 250,000 hours exposure for 18Cr steels of various grain sizes (4).

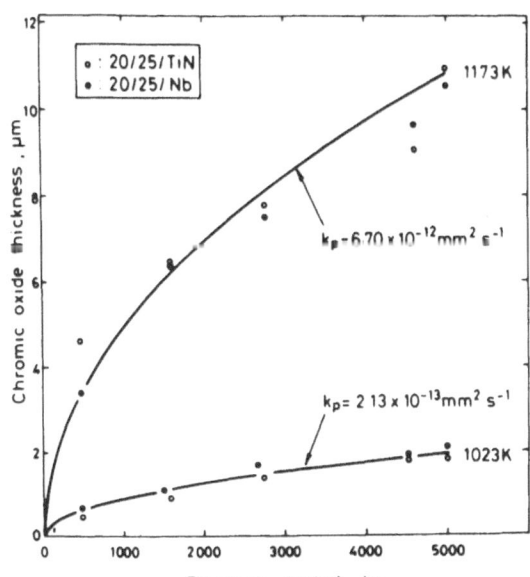

Figure 5: Growth kinetics of the protective oxide layer on 20Cr steels. Solid lines are the best-fit parabola (9).

In the graph:
- o : 20/25/TiN
- • : 20/25/Nb

Y-axis: Chromic oxide thickness, μm

X-axis: Exposure period, hr

$k_p = 6.70 \times 10^{-12}\,mm^2\,s^{-1}$ — 1173K

$k_p = 2.13 \times 10^{-13}\,mm^2\,s^{-1}$ — 1023K

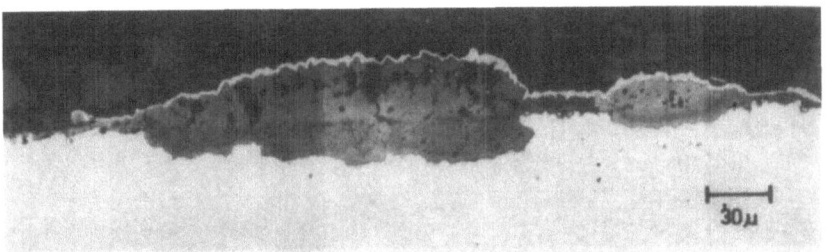

30μ

Figure 6: Localised pitting attack in a 20Cr/25Ni/Nb-stabilised steel oxidised for 6000 hours at 1123K (11).

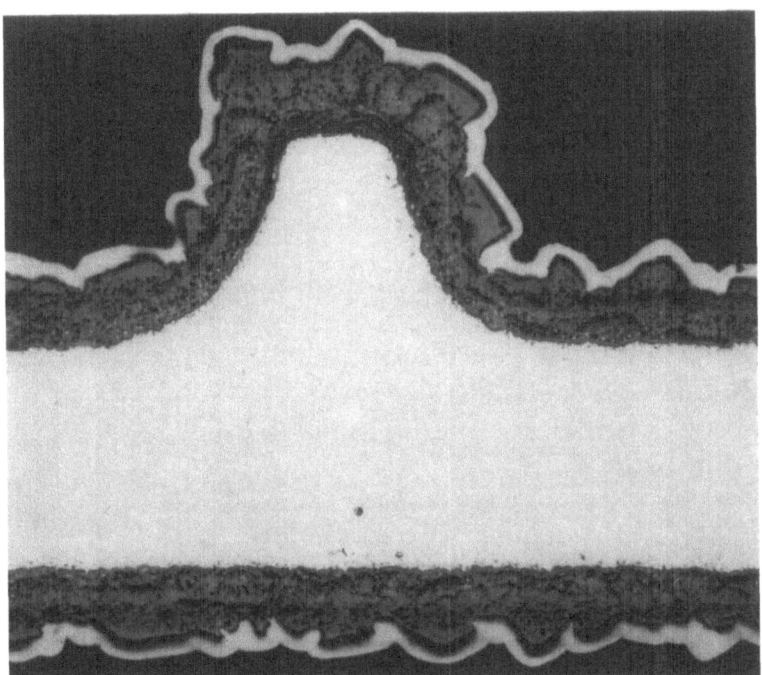

Figure 7: Section of unirradiated 20Cr/25Ni/Nb fuel cladding heated to 1473K in 15 minutes and held at temperature for 1 hour (15). This shows the residual metal under a porous oxide covered by a reflective Ni coat deposited prior to specimen preparation.

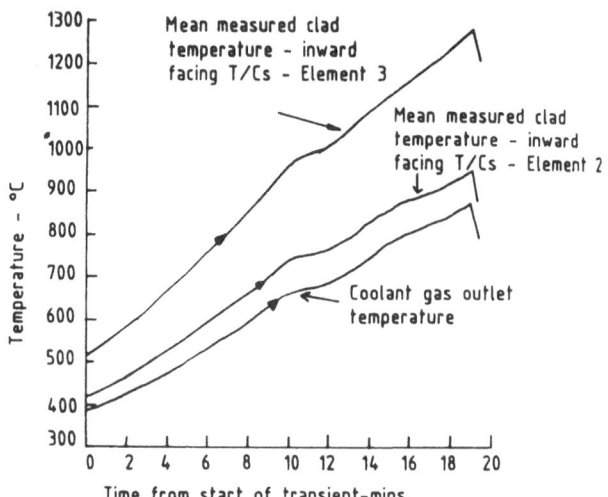

Figure 8: The temperature transient experienced by fuel cladding during 'Experiment 1' of the concluding experiments in the Windscale AGR loop (18).

24

5 Corrosion Behaviour of Metallic Materials in the Cooling Gas of High Temperature Reactors

W. J. Quadakkers and H. Schuster

The authors are in Kernforschungsanlage Jülich,
Institut für Reaktorwerkstoffe, P.O. Box 19 13,
5170 Jülich, FRG

SYNOPSIS

The reactive impurities in the primary cooling
helium of advanced high temperature gas cooled
reactors (HTGR) can cause oxidation, carburization or decarburization of the heat exchanging
metallic components. By studies of the
fundamental aspects of the corrosion mechanisms
it became possible to define operating
conditions under which the metallic construction
materials show, from the viewpoint of technical
application, acceptable corrosion behaviour. By
extensive test programmes with exposure times of
up to 30.000 hours, a data base has been
obtained which allows a reliable extrapolation
of the corrosion effects up to the envisaged
service lives of the heat exchanging components.

INTRODUCTION

In advanced HTGR's the primary cooling helium
transfers the nuclear heat from the graphitic
nuclear core to heat exchanging components
fabricated from metallic materials. In an HTGR
designed to supply heat for chemical processes
e g. coal gasification or steam reforming, these
components are operating at temperatures up to
950 °C. The coolant gas contains small amounts
of impurities, i.e. H_2, H_2O, CH_4, CO, N_2 and CO_2
in the Pa range which can react with the
metallic components thereby influencing the
mechanical properties of the alloys.

Extensive test programmes have been carried out
to investigate the corrosion behaviour of the
metallic materials in the different service
environments in order to obtain the data which
can be extrapolated to the envisaged service
lives of the components (more than 100,000 h).

MATERIALS AND TEST GASES

The selection of the candidate alloys for high
temperature heat exchangers (INCONEL 617,
HASTELLOY X, NIMONIC 86, Alloy 800 H, Thermon
4972) in HTGR plants was mainly based on long
term creep rupture properties, structural
stability and fabricability /1/. These Ni- and
FeNi- based alloys contain Mo, Co and Cr as
solid solution strengthening elements whereas
for the corrosion behaviour Cr, Mo and minor
alloying additions, as Mn, Ti, Al and Si, are
important.

As the investigations concerning the corrosion
behaviour of these alloys in simulated HTGR
helium were being carried out in different
laboratories, a standardized test gas, "PNP
helium" [1], was defined which contains 0.15 Pa
H_2O, 50 Pa H_2, 2 Pa CH_4, 1.5 Pa CO and less than
0.5 Pa N_2.

Although most of the testing has been done in
this standardized test gas, considerable effort
has additionally been devoted to the investigation of the influence of systematic variations
of the gas composition on the corrosion
behaviour.

----

1) PNP : Prototype Plant for Nuclear Process Heat

THEORETICAL BACKGROUND

The main corrosion effects occurring in HTGR helium are scale formation, depletion of the oxide forming elements, internal oxidation and carburization or decarburization. From the viewpoint of the influence on mechanical properties especially the changes in alloy carbon content are of importance.

In interpreting the experimental results one has to consider that due to the high gas flow rates in the primary circuit the corrosion behaviour of the metallic materials in a real plant will be determined by the kinetics of the individual gas-metal reactions and therefore a steady state rather than thermodynamic equilibrium will be established. Due to the slow kinetics of the carbon transfer by methane

$$CH_4 \rightleftharpoons C + 2H_2 \qquad (1)$$

the steady state potentials are very near to the equilibrium oxygen partial pressure and carbon activity of the reactions

$$H_2O \rightleftharpoons H_2 + 1/2 \ O_2 \qquad (2)$$

and

$$CO + H_2 \rightleftharpoons C + H_2O \qquad (3)$$

Only if the water level is extremely low (≈ 0.01 Pa), the corrosion behaviour is significantly influenced by methane splitting.

In respect to corrosion chromium is the main oxide and carbide forming alloying element in the candidate materials. Based on the classical thermodynamic stability diagram for this element a modified stability diagram for each candidate material can experimentally be determined for describing the corrosion behaviour /2/. The diagram can be divided into five areas in which different corrosion behaviour occurs, as shown schematically in Fig. 1. Atmospheres located in areas I or II lead to rapid decarburization without or with oxidation. In area IV and especially area V, rapid carburization occurs which causes a significant ductility loss of the materials at low temperatures ($\leqslant 700$ °C). Only atmospheres located in area III will cause no significant damage of the materials: irrespective of the steady state carbon activity in the gas, carburization occurs, but the stable chromia scale restricts the carbon uptake to below technically relevant levels /2/.

The borderline P_{CO}^* in Fig. 1 represents the equilibrium carbon monoxide pressure of the reaction

$$xCO + yMe \rightleftharpoons Me_yO_x + xC \qquad (4)$$

in which C represents the activity of carbon in the alloy and Me_yO_x the scale forming oxide. Which of the above mentioned corrosion effects will occur is mainly determined by the value of P_{CO}^*: atmospheres with a carbon monoxide partial pressure higher than P_{CO}^* are located in area III or IV; if the carbon monoxide pressure is lower than P_{CO}^*, the atmosphere is located in area I, II or V. Decreasing the water level or strongly increasing the methane partial pressure at constant CO level shifts the location of the atmosphere from area II to V or from area III to IV.

The value of P_{CO}^* depends on alloy composition and increases with temperature. As can be deduced from eq. (4), P_{CO}^* is decreased by addition of alloying elements which increase chromium activity, or/and increase the stability of the carbides or the scale forming oxides. Due to a lack of thermodynamic data for the complex oxides and carbides formed, P_{CO}^* cannot be calculated for the technical alloys from standard thermodynamic data. Therefore a technique has been developed by which P_{CO}^* could be determined experimentally /2, 3/: for the high temperature alloys under consideration it equals around 0.1 Pa at 800 °C and around 3 Pa at 950 °C /2/.

As P_{CO}^* depends on the activities of the oxide forming element(s), in practice P_{CO}^* is not completely constant. Because element depletion occurs in the near surface region due to scale formation, P_{CO}^* increases with time. Therefore the boundary between area II and III respectively IV and V in Figure 1 is not exactly determined by P_{CO}^* /2/. This is indicated by the shaded areas in Figure 1. An atmosphere located in these areas initially leads to a corrosion effect typical for area III (or IV), however after longer times it changes to a behaviour typical for area II (or V). It has been shown that for the conditions considered here the change of P_{CO}^* with time is not very significant /2/.

GAS COMPOSITION AND TEMPERATURE DEPENDENCE OF CORROSION BEHAVIOUR

To illustrate how the modified stability diagram (Fig. 1) can be used for the interpretation and prediction of the corrosion results, Fig. 2 shows the change in carbon content of the main reference alloy, INCONEL 617, during exposure at 850 and 950 °C in PNP helium and in a similar test gas, however with an extremely low water level.
The value of P_{CO}^* for INCONEL 617 at 850 °C is around 0.1 Pa /2,3/ which is significantly lower than the carbon monoxide pressure of 1.5 Pa in the PNP helium. Due to the water partial pressure of around 0.1 Pa chromium oxide is stable, and so at 850 °C PNP helium is located in area III of the stability diagram resulting in the formation of a pure oxide scale (Fig. 3a) which restricts the change in carbon content to very low levels.

This behaviour is found for all the candidate alloys in the temperature range up to 900 °C. At temperatures above around 920 °C (depending on alloy composition), however, P_{CO}^* exceeds the carbon monoxide pressure in the test gas shifting the atmosphere to area II of the stability diagram resulting in a very rapid decarburization /2, 3, 4/ with deleterious consequences for creep strength /5/ as illustrated by Figure 4.

In the helium composition with the very low water partial pressure the low steady state oxygen partial pressure causes the carbide to be the stable phase instead of the oxide. Therefore at 850 °C the atmosphere is located in area IV resulting in the formation of mixed oxide carbide scales (Fig. 3c). At 950 °C this dry atmosphere is located in area V causing the formation of a pure carbide scale accompanied by an approximately linear carbon uptake (Fig. 2b).

The results illustrate that at 950 °C in both atmospheres a rapid transfer of carbon occurs: rapid carburization in the dry atmosphere and rapid decarburization in the gas with 0,15 Pa water vapour. The reason is that at this temperature the carbon monoxide level in both gases is lower than P_{CO}^* and so the equilibrium of reaction (4) on the left side. Because of the extremely rapid kinetics of this reaction, chromium oxide and carbide (at the surface or in the alloy) cannot coexist. In atmospheres with a water level of around 0.1 Pa (area II) in which the oxidation by water dominates the carbon transfer by methane, the oxide scale which develops does not hamper carbon transfer because the oxide is continuously attacked at the scale-metal interface by the carbon in the alloy (reaction 4) leading to a destruction of the scale.

It could be shown /6/ that in a growing oxide scale always sufficient pores and cracks are initiated through which the carbon monoxide formed by reaction (4) can escape. The healing rate of the scale imperfections is, due to the low water partial pressure of around 0.1 Pa, too low to compensate the scale destruction caused by reaction (4). This destruction of the scale also occurs if the surface would be preoxidized /6/. In fact an undisturbed growth of the oxide in this atmosphere is only possible after the carbon loss has decreased the carbon activity to such a low level that reaction (4) is in equilibrium.

A stable behaviour which is characterized by a slight carbon uptake, like that shown in Fig. 2 for 850 °C in PNP Standard helium, not only requires the formation of a chromia surface scale but also a carbon monoxide level which is higher than P_{CO}^*. If these conditions are fulfilled, a stable corrosion behaviour is obtained also at the highest envisaged test temperature of 950 °C.

This is illustrated in Fig. 5 for INCONEL 617 where the corrosion behaviour in PNP helium (P_{CO} = 1.5 Pa) is compared with the behaviour in a test gas containing 10 Pa carbon monoxide which is higher than the P_{CO}^*-value for this material at 950 °C ($\approx$ 3 Pa). It can be seen that in the low-CO gas rapid decarburization accurs in spite of the higher methane level. In the test gas containing 10 Pa of CO, only a very slight change in carbon content occurs. A typical micro-structure after exposure in this test gas is given in Fig. 6b. Apart from the chromia based surface scale and internal aluminium oxide particles, a small carbide depleted zone is visible. Different from Fig. 6a where the carbide free zone is around 1 mm wide, the carbide depleted zone is not caused by carbon loss but by consumption of chromium used for scale formation. The width of this zone increases with time according to an approximately parabolic time law (Fig. 7), whereas the width of a carbide free zone caused by decarburization obeys an approximately linear time dependance.

CONCLUSIONS

The aim of the corrosion investigations accomplished by the partners in the German/Swiss HTGR project for nuclear process heat has been the determination of corrosion data which can be extrapolated to the envisaged service lives of around 100 000 hours. The majority of the experiments has been carried out in a standardized test gas, PNP helium. In the whole temperature range up to 900 °C this test gas is located in area III of the stability diagram determined for the main candidate material INCONEL 617 and so a stable corrosion behaviour, characterized by an oxide scale and very slight carburization, occurs. It has been found that, despite clear differences in corrosion behaviour between the different alloys and even between different heats of one alloy, all reference materials show this stable corrosion behaviour at and below 900 °C. Data compilations of the extensive test programmes, in which exposure times of up to 30 000 hours have been reached, show that the data base available at the moment enables the derivation of time laws necessary for reliable extrapolations of the corrosion effects up to the envisaged service live of around 100 000 hours.

At temperatures of around 950 °C, in the gas composition PNP helium the corrosion behaviour changes to a rapid transfer of carbon between alloy and atmosphere. However, based on the evaluation of the corrosion mechanisms the range of gas compositions could be defined in which even at the highest service temperature envisaged (950 °C) stable corrosion behaviour can be maintained for all candidate materials.

REFERENCES

/1/ H. Nickel, T. Kondo. P.L. Rittenhouse; Nuclear Technology 66 (1984) 12

/2/ W.J. Quadakkers, H. Schuster; Werkstoffe und Korrosion 36 (1985) 141 and 36 (1985) 335

/3/ L.W. Graham; High Temperature Technology; 3 (1985) 3

/4/ H. Brandis, B. Huchtemann, P. Schüler, H. Weber, W. Schendler; Stahl und Eisen 102 (1982) 417

/5/ R.V.D.Gracht,P.J.Ennis,A.Czyrska-Filemonowicz, H Schuster,H.Nickel; Proc. Int. Conf. on Creep, April 14-17, Tokyo 1986, p. 123

/6/ W.J.Quadakkers; Materials Science and Engineering 87 (1987) 107

ACKNOWLEDGEMENTS

The investigations presented have been carried out for the German HTGR project PNP in cooperation with the industrial project partners and subcontractors under the sponsorship of the Minister für Wirtschaft, Mittelstand und Verkehr des Landes Nordrhein-Westfalen and the Bundesminister für Forschung und Technologie.

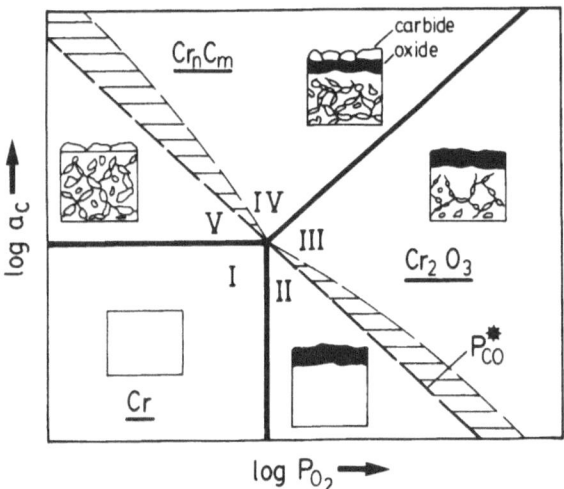

Figure 1: Corrosion areas in the modified stability diagram for chromium

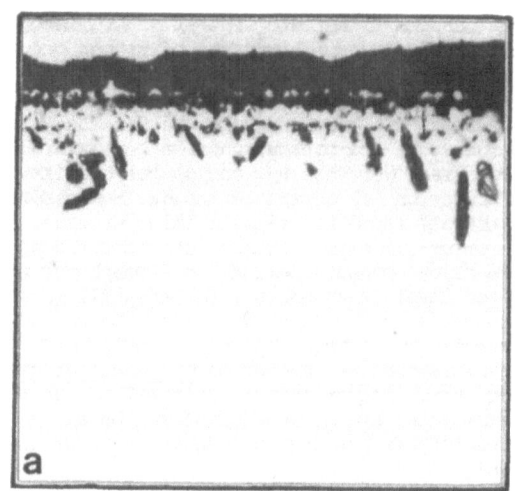

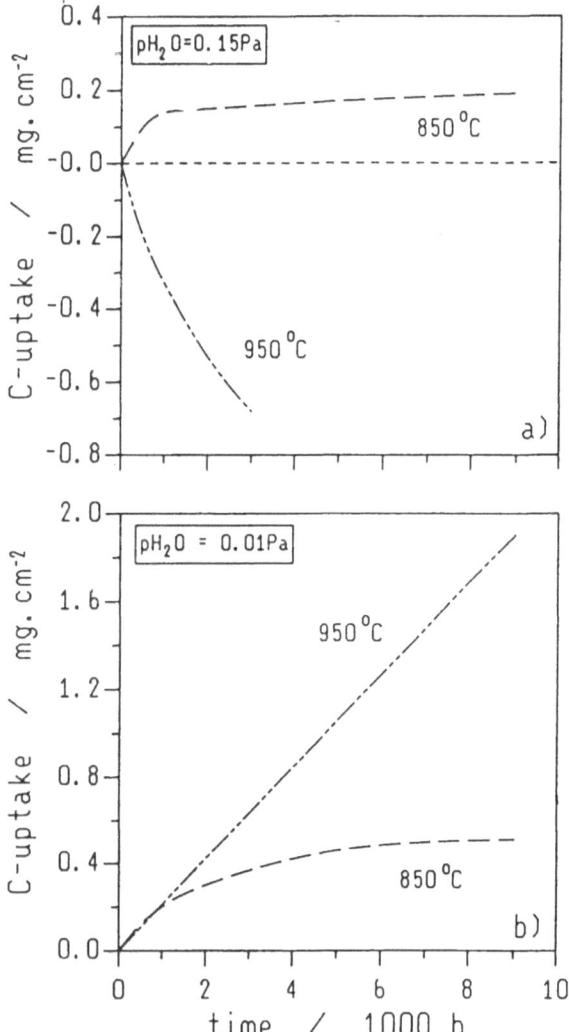

Fig. 2: Change in carbon content of INCONEL 617 during exposure in "PNP helium" with (a) 0.15 Pa and (b) 0.01 Pa H_2O

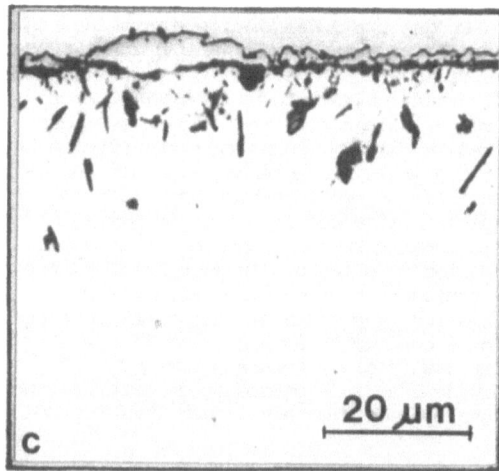

Fig. 3: Scale formation of INCONEL 617 in (a) PNP helium at 850 °C, 4000 h, (b) PNP helium at 950 °C, 2000 h and (c) PNP helium with 0.01 Pa H_2O at 850 °C, 4000 h.

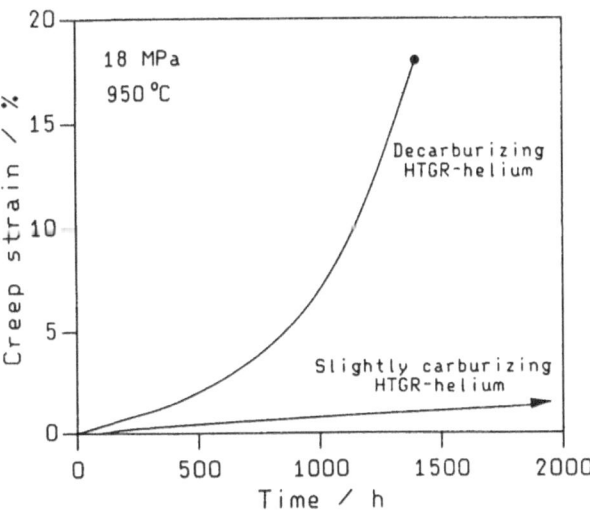

Fig. 4: Creep curves for alloy INCONEL 617 in different environments showing the deleterious effect of decarburization on creep strength /5/.

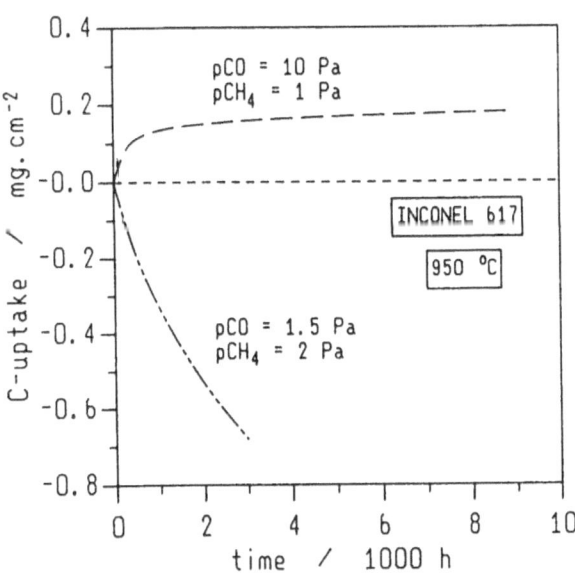

Fig. 5: Influence of CO partial pressure on the change in carbon content of INCONEL 617 at 950 °C. Testgas helium containing (in Pa) 0.15 H_2O, 50 H_2 and different concentrations of CO and CH_4

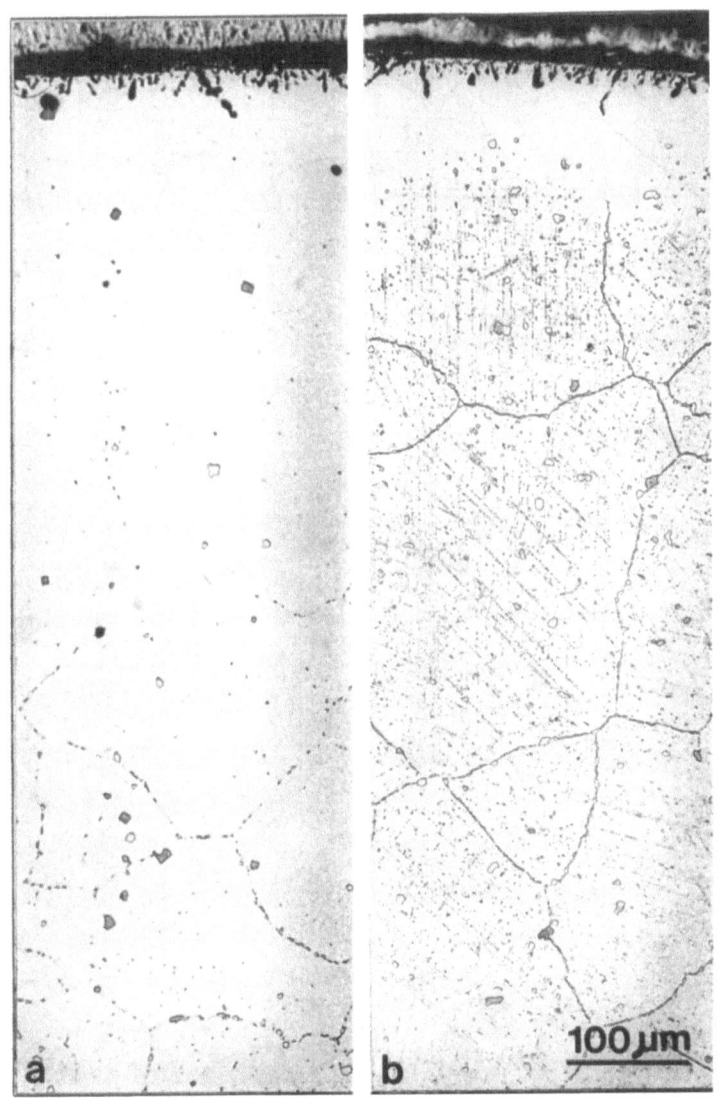

Fig. 6: Microstructure of INCONEL 617 after 1000 h esposure at 950 °C in HTGR helium compositions given in figure 5.
a) P_{CO} = 1.5 Pa, P_{CH_4}= 2 Pa; b) P_{CO} = 10 Pa, P_{CH_4} = 1 Pa

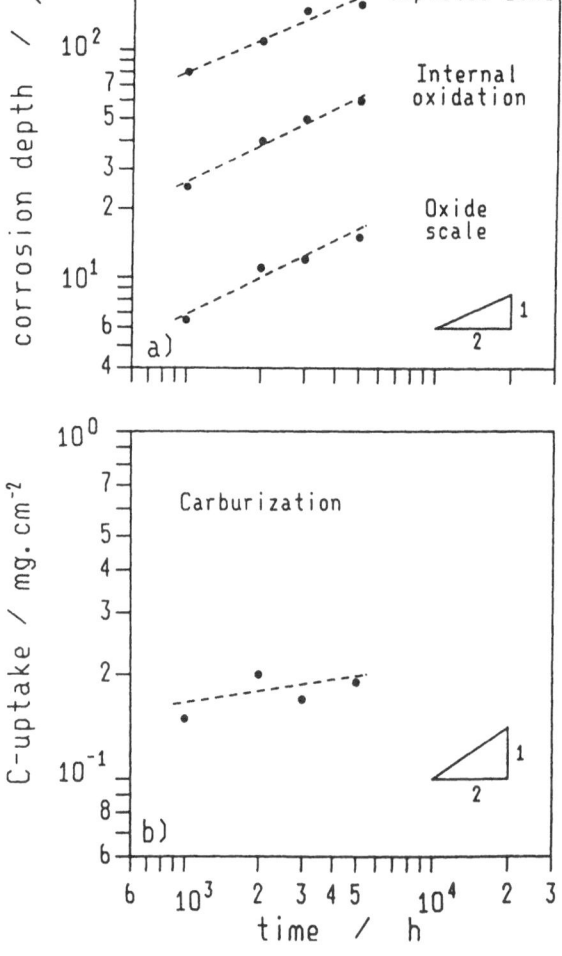

Fig. 7: Corrosion kinetics of INCONEL 617 at
950 °C in He-H_2O/H_2/CH_4/CO-0.15/50/1/10 Pa

6 Corrosion in Fast Breeder Reactors

H. U. Borgstedt and L. Champeix

H. U. Borgstedt is in the Institute of Materials
and Solid State Research, Kernforschungszentrum
Karlsruhe, Fed. Rep. of Germany;
L. Champeix is head of the Liquid Metals
Division, Centre d'Etudes Nucléaires de
Cadarache, France

1. INTRODUCTION

Breeder reactors need fast neutrons for breeding
and fission. Water is not suitable for the
cooling of the fuel elements. The high density of
energy requires a coolant with a very good
thermal conductivity. Liquid sodium is such a
medium, and it is liquid between 100-1000°C at a
pressure below four bars. Its viscosity is
comparable to that of water and its compatibility
with several materials is fairly satisfactory. Its
neutron-physical properties lead to relative high
breeding ratio. The activation of sodium
according to

$$^{23}Na(n, \gamma)~^{24}Na~\text{or}~^{23}Na(n, 2n)~^{22}Na$$

requires shielding of the primary cooling
system. The hazards caused by sodium-water
reactions in a sodium-heated steam generator are
separated from the activated primary system by
the use of a secondary sodium system transferring
the heat generated in the core via a sodium heat
exchanger to the steam generator or superheater
of the water circuit.

Though there are no hazards that activated
sodium might be exhausted into the atmosphere by
means of a violent sodium-water reaction, there
are still problems to be solved which are due to
the material optimization in the steam generator
and the secondary sodium circuit itself. Thus,
the primary and secondary systems have their
specific corrosion problems which are dependent
on the temperature and flow velocity of the
sodium, the choice of materials and their
combination in a cooling circuit and of chemical
parameters.

2. CORROSION CAUSED BY FLOWING SODIUM

Corrosion by flowing sodium is caused by the
solubility of metals in the liquid metal or by
chemical reactions of constituents of materials
with impurities in the sodium. The reactions may
cause the formation of more or less adherent
layers or the loss of the reaction products due
to solution or spalling off by the flowing
liquid. Alloys as autenitic CrNi steels are
corroded in both ways. The combined processes of
dissolution and chemical reactions create typical
corrosion phenomena. Phenomena and corrosion
rates are dependent on parameters as temperature,
flow velocity, down-stream position, purity of
sodium in respect to oxygen and carbon, and the
presence of materials acting as sinks for
dissolved elements.

2.1. CORROSION OF CORE COMPONENTS AND FUEL-
ELEMENT CLADS

Since the core components - fuel-element clads,
wrapper tubes, distance holders, etc. - are
exposed to the coolant at the highest temperature
level, their materials tend to lose constituents
to the flowing sodium. The loss of material
depends on the temperature according to equation
(1) valid for a cladding tube at down-stream
position zero and a sodium velocity of 3-5 m/s
[1]

$$log_{10} S[\mu m/a] = 3{,}845 + 1{,}5 \, log \, [0] - \frac{18000}{4{,}604 \cdot \; T} \qquad (1)$$

Equation (1) is developed for stainless steel AISI 316 or 304; it has also been proved for several stabilized austenitic steels [2]. The loss of wall thickness S is calculated from weight loss measurements. The chemical attack of sodium is, however, not uniformly distributed. A part of the weight loss is caused by the selective leaching of alloying elements. Thus, a small loss of thickness occurs. Grain boundary grooving indicates faster dissolution processes at these parts of the surfaces, and the surface layers are changed in their chemical composition due to losses of chromium, nickel, and manganese [1-4]. The chemical changes may lead to structural changes, ferritic zones at the surface - and grain boundaries are created. Fig. 1 shows an example of a cladding tube of steel X8 CrNiMoVNb 16 13 exposed 5000 h to flowing sodium at 700°C [5].

The formation of cavities at triple grain boundaries is observed in surface layers of some of the steels. Thus, one has to conclude that the ferrite layer does not contribute to the strength of the tubes.

The steels tend to exchange carbon with the liquid metal. At 700°C the diffusion rate of carbon in steel is fast enough to cause changes of carbon concentrations in significant parts of the cladding tubes. While unstabilized steels are decarburized at this high temperature, stabilized steels pick-up carbon from sodium. The carburization affects the whole wall thickness of cladding tubes as is demonstrated in Fig. 2 [5].

The uptake of carbon influences the microstructure, since carbides precipitate in the region of the highest concentrations. On the inner side, where carburization remains limited, the σ-phase occurs after long exposure at 700°C. The changes of carbon concentrations are important for the mechanical properties of the cladding tubes, carburization increases the strength and reduces the ductility of the steels. High carbon contents are detected by means of microhardness measurements.

An aspect related to safety and reliability of the system is the leaching of activated elements in corrosion processes, thus activating the cooling systems. Such active isotopes are 51Cr, 60Co, 54Mn, and 58Co. While chromium and manganese have the tendency to migrate through the sodium circuits (from the core to the heat exchanger), cobalt remains in the high-temperature region. Activation of components (pumps, heat exchanger, etc.) rises costs of the maintenance and the shielding of the primary system.

2.2   CORROSION OF STRUCTURAL MATERIALS

The materials of the tank and primary piping of a fast reactor are exposed to sodium at a lower level of the temperature than the core components. Their position is down-stream to the temperature maximum at the core exit. Thus, dissolution and loss of material does not seem to be of concern for the life time of the components. Many regions of the system are effected by material deposition. The deposits cover the surfaces of the tubes, since they form dense surface layers of non-metallic (oxidic or carbidic) material. Chromium is enriched in these layers. They also contain some crystals of metallic appearance, in which nickel and manganese are enriched. The constituents of these crystals tend to diffuse into the matrix of the structural material.

Carbon exchange seems to be of lower importance at the lower temperature level of the primary tubing of a fast reactor. The diffusion being much slower does not affect very thick zones of the tube walls. An effect of carburization on the creep-rupture behaviour of the steel AISI 304 is observed. As far as it is only shortening the tertiary creep region, it is less important for the creep life time of components. Decarburization of the structural material may cause a slight reduction of its high-temperature strength [6]. Not all steels are sensitive to such effects. It does not seem that sodium of 550°C influences the fatigue life of stainless steels, though some effects on the fracture mode are known. The exchange of nitrogen between steels and sodium generates similar effects as does the carbon exchange.

2.3   CORROSION OF STEAM GENERATOR MATERIALS

Materials to be used in steam generators are the austenitic alloy X 10 NiCrAlTi 32 20 (Incoloy 800) and ferritic steels as 2,25 Cr 1Mo since these materials are less sensitive to local attack from the water side [7]. While the austenite has a higher Ni content than the structural steels AISI 304 or AISI 316, the ferritic steels do not contain considerable amounts of nickel. Thus, the use of these materials can give rise for some mass transfer of nickel. Incoloy 800 acts as a source for nickel in the sodium circuit, the ferritic steels as sinks.

The steam generators are, however, the heat sinks in the secondary sodium circuit. Thus, they should rather act as traps for dissolved materials than as components to be corroded by the flowing sodium. Incoloy 800 may partly suffer some depletion of nickel, the effects, however, should be very limited. It has been found that at temperatures ≤ 600°C the formation of internal iffusion layers can nearly be neglected [8], the losses of material are lower than 100 g/m²·a or 10 μm/a [7].

Though carbon transfer between ferritic 2,25Cr 1Mo steel and austenitic steel 316 L is favored by thermodynamics, no loss of carbon of the ferrite occurs up to a temperature of 475°C, which is the maximal temperature in the steam generator of the PHENIX reactor [9]. Ferritic steels with higher chromium contents (up to 9 wt-%) do not suffer decarburization even at higher temperatures.

2.4   CORROSION UNDER ACCIDENT CONDITIONS

Accidents may cause a rise of the temperature in the core. Thus, the materials of the core components may be exposed to sodium at higher than the normal temperature level. Corrosion rates are known to be temperature dependent according to equation (1). Experimental measurements have been made up to a temperature of 760°C. The equation may be used to this temperature, and extrapolated up to the boiling temperature of sodium (~ 880°C) Heating of the sodium to 760 resp. 880°C would increase the corrosion rates to 0,01 resp. 0,03 μm/d. The effect of such excursions of the operation temperature is negligible since the duration of such events has to be considered to be short.

The increase in temperature may however, generate high stress on the cladding tubes. This high stress is not harmful if inealstic deformation of the material does not occur. If it occurs, a limited sodium effect might be observed at surface grain boundaries. The effects need some time to develop deterioration of the materials, since all these deteriorations are based on diffusion processes.

Accidents may increase the impurity levels of the coolant by leakage, which allows water or air to enter the system. Sodium hydroxide is found to cause stress corrosion of austenitic steel AISI 316 L, if the concentration of NaOH, the temperature and the stress are high enough [10]. A more detailed study has been made on the behaviour of Incoloy 800 in sodium-sodium hydroxide mixtures under applied stresses[7]. As is indicated in Fig. 3, the occurrence of stress corrosion depends on the hydroxide concentration and the stress level. If a certain stress at a given concentration of hydroxide is exceeded, intergranular stress corrosion of the material is observed. At a temperature below 420°C, hydroxide is not stable in contact with liquid sodium. In the lower temperature region the effects are therefore less pronounced.

3. CONCLUSIONS

Sodium corrosion effects are small compared to effects in gaseous circuits operated at high temperatures. It is, however, necessary to keep the conditions of operation within a range such that corrosion is limited to a low level to prevent an activation of the system with long life radioactive corrosion products.

Several materials are suitable for use in liquid sodium reactors, among them ferritic and austenitic and high temperature alloys with up to 32 wt% nickel contents. The designers have, however, to consider the mass transfer between materials of different compositions. The exchange and transfer of non-metallic elements, such as carbon or nitrogen, has to be taken into account.

Safety and reliability of sodium cooled reactors are almost unaffected by sodium corrosion. Leakages may occur if a component or tube contains weak points at which stresses are concentrated or weldments are not properly performed. Thus, the design and construction has to be checked thoroughly.

Mass transfer causes deposition or precipitation of material. It has never been observed that such precipitations have been able to block a sodium system. Precipitated layers, however, decrease the heat conduction and increase the pressure drop of tubes or channels. Ferritic surface layers, on the other hand, interfere with magnetic fields of flow meters and pumps.

Some more severe sodium effects may occur after accidents which cause increases of the temperature and the impurity contents of the liquid metal. They may be limited by a restriction of the unusual parameters to short periods of time.

REFERENCES

[1] A.W. Thorley and C. Tyzack, in: Liquid Alkali Metals, The British Nuclear Energy Soc., London 1983, 257-274

[2] H.U. Borgstedt, in: 2nd Internat. Conf. on Liquid Metal Technology in Energy Production, J.M. Dahlke, Ed., NTIS, Springfield, Va., 1980, 7,1-10

[3] P.A. Baque, L.J. Champeix and E.T. Honnorat, in: Chemical Aspects of Corrosion and Mass Transfer in Liquid Sodium, Ed. Sa.A. Jansson, The Metallurgical Soc. of AIME, New York 1973, 299-316

[4] P. Baque, L. Champeix, A. Lafon and E. Sermet, in: Liquid Alkali Metals, The British Nuclear Energy Soc., London 1973, 223-231

[5] H.U. Borgstedt and W. Dietz, Report KfK 2516, Kernforschungszentrum Karlsruhe, 1977

[6] H. Huthmann, G. Menken, H.U. Borgstedt and H. Tas, in: 2nd Internat. Conf. on Liquid Metal Technology in Energy Production, J.M. Dahlke (Ed.), NTIS, Springfield, Va., 1980, Vol. 2, Chapter 19, p. 33

[7] L. Champeix, in: ALLOY 800, W. Betteridge et al., eds., North Holland Publ. Comp., Amsterdam 1978, 283-289

[8] H.U. Borgstedt, G. Frees, A. Marin, in: ALLOY 800, W. Betteridge et al., eds., North Holland Publ. Comp., Amsterdam 1978, 291-295

[9] P. Baque, M. Besson, L. Champeix, J.R. Donati, C. Oberlin and P. Saint-Paul, in: Internat. Conf. on Liquid Metal Technology in Energy Production, M.H. Cooper (Ed.), NTIS, Springfield, Va., 1976, Vol. 2, 834-840

[10] L. Champeix, P. Baque, and C. Chairat, in: 2nd Internat. Conf. on Liquid Metal Technology in Energy Production, J.M. Dahlke (Ed.), NTIS, Springfield, Va., 1980, Vol. 1, Chapter 8, p. 49

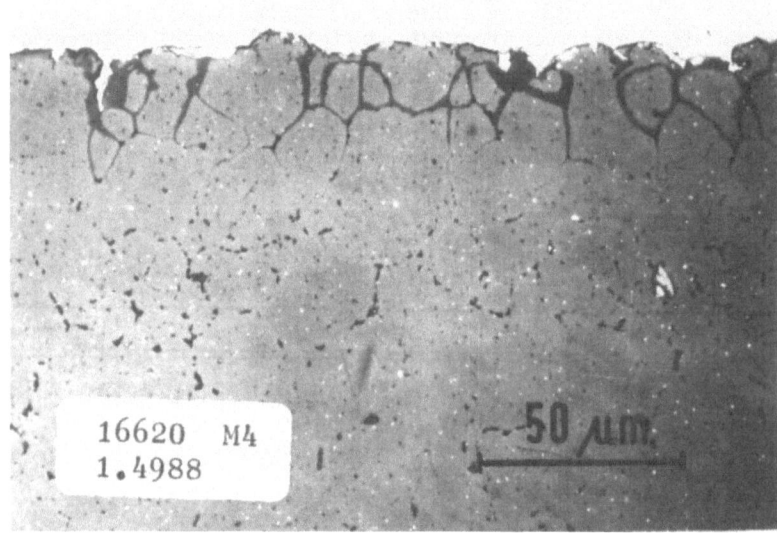

Fig. 1

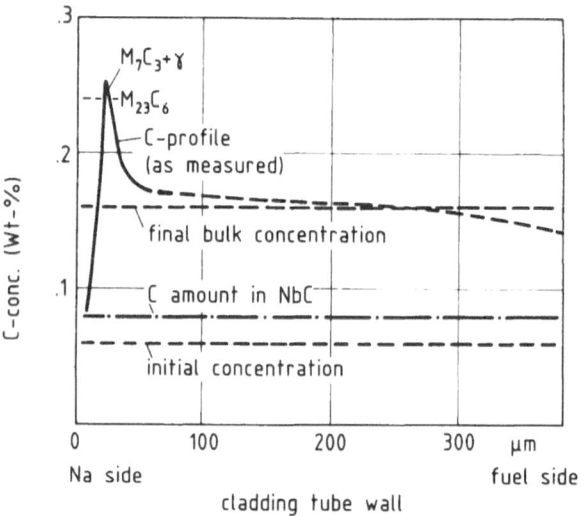

Fig. 2: Carburization of a cladding tube of steel X8 CrNiMiNb 16 16 in sodium of 700°C after an exposure of 5000 hrs

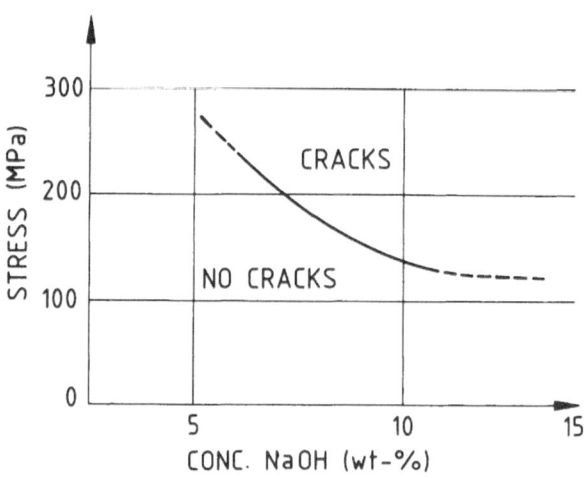

Fig. 3: Sensitivity to intergranular crack corrosion of Incoloy 800 in Na/NaOH mixtures

7 Corrosion Problems in Nuclear Fusion Reactors

V. Coen

Dr. Coen is with the Materials Science Division of the Joint Research Centre of the Commission of the European Communities at Ispra (VA) Italy.

SYNOPSIS

In order to produce Tritium by neutron irradiation in a fusion reactor the use of Li or of Lithium bearing material is essential.
The Breeding materials most commonly considered are liquid metals or alloys (Li, Pb-17Li) and ceramics (Li$_2$0, LiAl0$_2$, Li$_2$Si0$_3$, Li$_4$Si0$_4$, Li$_2$Zr0$_3$).
The coolants are H$_2$0, the liquid metals or alloys (self cooled concepts) or He. The structural materials envisaged comprise austenitic stainless steels (AISI 316, Cr-Mn steels), ferritic or martensitic steels (HT9, 1.4914).
In this paper the main corrosion/compatibility effects are described and the results obtained critically reviewed. Special emphasis is placed on the compatibility between liquid metal or alloy and structural materials relevant to NET (Next European Torus).

1. INTRODUCTION

In nuclear fusion nuclei of two light elements join together to produce a heavier nuclei with a contemporaneous emission of energy.
While many fusion reactions are possible the one envisaged for probable utilization in the near future is the D-T reaction

$$D + T \longrightarrow He^4 + n + 17.58 \text{ MeV} \qquad (1)$$

T has a half life of only 12.3 years, it is not found naturally in useful concentrations and must be produced.
Lithium is the breeding material for T, it can absorb a neutron and produce T and He according to the following reactions:

$$Li^6 + n \longrightarrow T + He^4 \qquad (2)$$

$$Li^7 + n \longrightarrow T + He^4 + n' \qquad (3)$$

If neutron multipliers as Be and Pb (n, 2n) are present the breeding ratio will be adequately increased.
In a fusion reactor (exemplified in Fig. 1) a D and T plasma is confined at a very high temperature ($\sim 10^8$ K) for a specific length of time; the neutrons resulting from the fusion reaction breed tritium and release their energy to the blanket producing heat that is conducted away via a coolant.
Two kinds of confinements are studied the magnetic (Tokomaks and Mirrors) and inertial.
Presently fusion reactors are still at the stage of Conceptual Design Studies and the Tokamak concept is the one most studied. This

is especially so in Europe, where the NET (Next European Torus) is the main fusion device which according, to the European fusion energy development strategy, will follow JET (Joint European Torus) and provide the basis for designing a DEMO (Demonstration Reactor).

In this paper we will concentrate on the materials compatibility problems related to the Tokamak type reactor. The first wall, as its name explicitly indicates, is the first barrier interfacing with the plasma; candidate structural materials for the first wall are: austenitic stainless steels and especially AISI 316 (on account of the large data base on the fission reactor performance of this alloy); ferritic (martensitic) steels such as HT-9 (Fe-12Cr-1Mo,VW), DIN 1.4914 (Fe-10.5Cr,MoNbV), Fe-9Cr-1Mo and the low activation Vanadium based alloys. The two NET reference materials are AISI 316 L and DIN 1.4914.

In direct contact with the first wall is the blanket which has to breed tritium for the fuel cycle, convert the fusion energy into heat and conduct that heat away to an energy producing system. The blanket consists of the structure, breeder and coolant. The candidate structural materials are the same as for the first wall. The breeders are, of course, liquid or solid lithium compounds. The most commonly proposed are for the liquids: pure Li and the eutectic Pb-17Li; for the solids: Li_2O, $LiAlO_2$, Li_2SiO_3, Li_2ZrO_3, Li_4ZrO_4.

The separate coolants are: H_2O or He. In the self cooled concepts, naturally, Li and Pb-17Li are also considered. To a lesser degree, the use of liquid Na and molten salts has also been envisaged.

The NET options are: (a) for the liquid breeder, Pb-17Li (H_2O cooled) and (b) for the solid breeder concept: all the above mentioned oxides (He cooled). Most of the corrosion compatibility problems encountered in a fusion reactor concern the blanket-coolant combinations. It will be attempted to examine them in detail, concentrating on the NET priorities.

It must however be stressed that, in none of the tests carried out, the severe fusion reactor environment (14 MeV neutron radiation and high magnetic field) is completely reproduced.

2. CORROSION CAUSED BY INTERACTION BETWEEN THE FIRST WALL/BLANKET STRUCTURAL MATERIAL AND LIQUID BREEDING/COOLANT MATERIALS.

2.1. General consideration of corrosion by liquid metals

Corrosion of materials by liquid metals may occur by different mechanisms: dissolution, mass transfer due to thermal gradients, interstitial element transfer and intergranular penetration. Usually, more than one of these, if not all of them, contribute to the deterio-

ration (wall thinning, flow restrictions, accumulation of radioactive material, loss of mechanical integrity) of the system studied, and this depends on parameters peculiar to each specific material-liquid metal system.

Important factors affecting the corrosion are, for instance:

- the purity of the liquid metal (interstitial content)
- the composition and microstructure of the containment material
- operating temperature
- time
- temperature gradient in the system
- flow velocity.

Of course, some of these parameters, especially the last two, depend on the use of the liquid metal. If it is used only as a tritium breeder, it can be semi-stagnant and thus with only small temperature gradients in the system. If it is also used as a coolant, high flow velocities and large ΔT will be present, with an increase in the mass transfer problems.

The corrosion tests which are usually carried out include:

- isothermal static tests (in capsules or pots) to determine the basic mechanisms of reaction between the components of the system including impurities and interstitials;
- the study of corrosion and mass transfer in non-isothermal systems. This is usually carried out in loops. In the case of semi-stagnant tritium breeding blankets, thermal convection loops (TCL) with low flow velocity and small temperature gradients are sufficient while where higher velocities and T are required, forced convection loops (FCL) are necessary.

2.2. Lithium

It is only during a limited number of years that lithium compatibility studies have been directed towards experimental conditions corresponding to the requirements of the fusion reactor conceptual designs. Most of the tests carried out previously have the disadvantage of dubious purity of the liquid metal as well of the handling procedure so that data are often contradictory and not easy to evaluate. Good review papers exist on the pre-fusion era of lithium technology /1,2,3,4/. Nearly all of the recent studies are mainly on austenitic steels (Ni or Mn rich) and to a much lesser degree on the ferritic-martensitic steels.

2.2.1. Austenitic stainless steels (AISI 316, 316 L, 304, Cr-Mn steels):

Two main phenomena characterize the action of Li on these steels: dissolution and grain boundary attack; the relative effects depend largely on the experimental conditions. Ni, Mn and to a lesser degree Cr are dissolved from the steels. The dissolution rates in flowing Li increase with an increase in the thermal gradient or flow velocity in the system. No systematic correlation is however available. As a result of Ni or Mn depletion a porous ferritic layer is developed on the surface of the steels. The thickness of the ferrite layer, which corresponds to the depth of penetration when no grain boundary attack is evident, or the weight loss are usually used to express the corrosion rates.

Nitrogen present as Li_3N in the Li or dissolved in the steel plays an important role in the corrosion phenomenon. It provokes a grain boundary attack and increases the dissolution rate of Cr. Both phenomena are linked to the formation of a Li-N-Cr compound. Grain boundary attack was first reported /5/ in AISI 316 at 1089 K (1-2 wt.% Li_3N) then in AISI 304 L (2 wt.% Li_3N) /6-7/ and in nitrogen saturated 304 L. The nucleation of grain boundary carbides preceding penetration was suggested. It was shown /8/ by microanalytical techniques that 316 exposed to Li, containing nitrogen at unit activity at 923 K was subject to substantial Li penetration along grain boundaries, (Fig. 2) and the formation of a surface corrosion product of the type Li_9CrN_5 was postulated. By studying the grain boundary penetration (Fig. 3-4) of Li in nitrogen alloyed Cr-Mn steels /9-10-11/ it was clearly shown that at 873 K the $M_{23}C_6$ carbides, formed prior to the Li penetration, rearrange and change their composition in the presence of Li. Coherent twin boundaries (in which carbides are not precipitated) are not penetrated by Li The presence of a Li-Cr-N compound corresponding probably to Li_9CrN_5 has also been evidenced in the grain boundaries by the SEM technique (Fig. 5). As far as dissolution is concerned it is reported /15/ that in an FCL (Forced Convection Loop) greater Cr depletion and weight loss were encountered in 316 with Li containing 250 ppm N than in Li with < 100 ppm N. On the other hand, it seems that in a TCL (Thermal Convection Loop) /3/ similar changes in Nitrogen content cause no effect on corrosion. It must be stressed that the experimental conditions may have been different (possibility of N gettering?). It is clear that additional data are required to establish the actual effect of nitrogen on corrosion and mass transfer in flowing Li systems. Since the intergranular penetration of Li and the formation of Li_9CrN_5 is apparently linked to carbide precipitation, it is possible that in the temperature range in which no carbide precipitation occurs for relatively long times this attack could be avoided. It has been shown /14/ that when the carbon content of the steel was lowered, by working in presence of a hydrogen partial pressure, no grain boundary penetration could be observed under the same experimental conditions in which it had previously been detected (Fig. 6-7).

As far as oxygen is concerned, it does not appear to have any influence on the attack of steels; the surface oxides are reduced by Li with formation of Li_2O. No increase in dissolution rates has been reported.

In Fig. 8 the main features of the corrosion mechanisms for the Li/austenitic stainless steel couple are schematically summarized.

2.2.2. Ferritic-Martensitic steels:

There is far less information on the mechanism of corrosion of these steels in lithium. In static tests /15/, with pure lithium, specimens of 2 $1/4$Cr-1Mo steel were decarburized whereas steels with higher chromium content (HT-9) did not appear to be decarburized. This decarburization phenomenon is strictly dependent on the exposure temperature and on heat treatment, so that it can certainly be minimized. In tests carried out in non-isothermal, flowing lithium it would seem /16/ that after a relatively large weight loss (associated with Cr depletion) during the initial 500 h exposure to Li the weight loss in HT-9 and Fe-9Cr-1Mo steels follows a linear behaviour with time and yields constant dissolution rates; the dissolution rates for HT-9 and Fe-9Cr-1Mo steels increase by a factor of 10 when the temperature increases from 673 to 823 K, and an increase in nitrogen in Li increases the dissolution rates. This, again, may be linked to the formation of the Li-Cr-N compound.

Weight change measurements indicate for 12 Cr - 1 Mo VW steel, exposed to thermally convective Li between 873 and 723 K for 7000 h, a dissolution rate of 3.0 $mg/m^2.h$.

In the temperature range 773-623 K the weight loss was only 0.7 $mg/m^2.h$. This large difference is explained /17/ by assuming that weight changes reflect the competition between the following phenomena: dissolution/deposition of steel constituent, soluble reaction product formation and soluble surface reaction products. At higher temperatures thermal gradient elemental transfer would tend to dominate such that a relatively large net weight loss might result at T max. Such temperatures may be above the temperature range of stability of the surface product or preferential dissolution of Cr at this higher temperature may significantly reduce its activity at the surface. As the temperature range of a loop experiment is lowered the relative importance of thermal gradient mass transfer decreases and corrosion product reaction can make a large contribution to the measured net weight changes and radically change the slope of the mass change profile.

2.2.3. Operating Limits for Steels in Li Systems:

Although the experimental data are not sufficient to clarify the influence of the various system parameters on the phenomena involved in the corrosion of steels by Li, a qualitative assessment of the operating temperature limits can be made by comparing data obtained in the different laboratories that have carried out experiments in flowing Li. Fig. 9 is a simplified version of the curve presented in /18/ to which data from C.E.N. Mol /19/ have been added and corresponds to an Arrhenius plot of corrosion rates (sound metal loss) for types 316 (including 304 values) and HT-9 in flowing Li.

It is clearly apparent that in FCL (representative of Li used also as the coolant) the dissolution rates for type 316 are an order of magnitude greater than in TCL systems (representative of Li used as breeding blanket only). Dissolution rates for the Ferritic steel HT-9 in TCL and FCL are identical and similar to 316 in TCL.

If the limits of 5 μm/y loss rates associated with radioactive transport and 20 μm/y associated with plugging are accepted /18/, it would appear that in circulating Li AISI 316 could be used in the range 723-773 K, while if Li is semi-stagnant it could be used in the temperature range 823-873 K. This upper temperature range would apply to HT-9 in both cases.

2.2.4. Mechanical Property Degradation:

Only limited information exists on the effect of Li on the mechanical properties of austenitic steels and even less on the ferritics. Most of the data come from static experiments. It would seem that Liquid Metal Embrittlement (L.M.E.) does not occur in the Li environment /20-21/ and that for austenitic stainless steels tensile and creep properties are not altered by the Li environment /20-22-23/. The creep strength of austenitic and ferritic steels, in flowing Li, is lower than in static Li or argon /16/. This is due to the formation of a weak ferrite layer on the austenitic steels and decarburization of ferritic steels. While fatigue life is not influenced by pure Li it is significantly reduced in flowing Li containing < 1000 p.p.m. N /18/ on account of intergranular attack of the material. Tests in flowing Li, of known purity are necessary to establish a clear relation between corrosion and degradation of mechanical properties.

2.3. Pb-17Li

It is only since about 1980 that the lithium lead eutectic containing 87 at% Pb and 17 at% Li (referred to herein as Pb-17Li) has appeared particularly interesting as an alternative to liquid Li /24/. The Li-Pb phase diagram is reported in Fig. 10. The immediately obvious advantage is the fact that breeder and neutron multiplier are combined in one and the same material. Moreover, the eutectic does not show vigorous chemical reactions with air or water which is an important criterion in the choice of breeder materials. Only a few data exist on the physical and chemical properties of Pb-17Li as well as on the corrosion and mass transfer behaviour of this alloy. The main reason is that up to now it had no commercial application (it is not even found in commerce). Most of the studies carried out in the recent years related to the structural material-breeding/coolant compatibility concern mainly AISI 316. In Europe the experiments were carried out in static and TCL conditions /19,25,26,27,28/ while in the U.S.A. an FCL was also operated and some data on the behaviour of ferritic steels (HT-9, Fe-9Cr-1Mo) are also available /18,29,30,31/. The corrosion mechanism of AISI 316 in Pb-17Li is characterized by a strong Ni depletion and penetration of Pb and Li into the matrix (Fig. 11) /25/. A porous ferrite layer is developed and in some cases it is very weak and spalls either during exposure to the lithium lead eutectic (especially in FCL tests) or during cleaning of the specimens. It is very difficult to remove the residual Pb-17Li from the corrosion samples. The different techniques adopted include dissolution in Li or Hg or Na+NH$_3$ followed by washing in alcohol and water. As far as the effects of interstitial, nitrogen and oxygen on the properties of Pb-17Li are concerned, it has been shown /32/ that while the former is absolutely inocuous, the latter is quite deleterious: it enhances the corrosion of AISI 316 and is able to extract Li from the eutectic alloy. The results for nitrogen were confirmed by a study /33/ in which it is reported that 316 steel is corroded by nitrogen in Pb-17Li only at pressures above 10^{12}Pa.

The mechanism of corrosion of AISI 316 in Pb-17Li in presence of oxygen was studied by carrying out tests with and without added oxygen /34/.

These experiments have shown that oxygen enhances the corrosion of 316 steel by increasing the depth of the ferritic corrosion layer (Fig. 12) and the extent of chromium depletion within the layer. Tests using lithium lead alloys with compositions between Pb-17Li and Pb-11Li and in pure Pb showed no change in the depth of the ferritic zone with composition of the alloys but a significant increase for pure lead. Measurements of the solubility of nickel in Pb-17Li and Pb-10Li showed the solubility to be similar for both compositions and to be much less than that previously measured in pure Pb. The solubility may be expressed by the equations:

$$\log_{10} S(\text{wppm}) = 4.832 - \frac{981.2}{T(K)} \quad \text{for Pb-17Li}$$

$$\log_{10} S(\text{wppm}) = 5.148 - \frac{1162}{T(K)} \quad \text{for Pb-10Li}$$

The increase in ferrite zone formation, in the presence of oxygen, cannot, therefore, be due to increased Ni solubility caused by a change in composition of the alloy. A study of the reactions of oxides with Pb-17Li has shown that Cr may interact with oxygen to form the ternary oxide $LiCrO_2$.
In contrast to Li no grain boundary attack has been evidenced in presence of Pb-17Li.
In Fig. 13 the main features of the corrosion mechanisms for the couple Pb-17Li/austenitic steels are schematically summarized.
Notwithstanding the paucity of the available results an Arrhenius plot of the up to date corrosion rates of AISI 316 (304) is presented in Fig. 14. Owing to the uncertainty of the weight changes. The values taken into consideration are those where the thickness of the ferritic layer is mentioned; where more than one value is given the highest has been taken.

2.3.1. Operating Limits for Steels in Pb-17Li

If, as in the case for Li the limits of 5 μm/y loss associated with radioactive transport and 20 μm/y associated with plugging are accepted the operating limits of AISI 316 in Pb-17Li would be in the range 623-673 K. The very few data on ferritic steels (HT-9, Fe-9Cr-1Mo) show, as in the case for Li, an increase in the temperature range of about 100 K.

2.3.2. Mechanical Property Degradation

The very few data available on the influence of the lithium-lead eutectic on the mechanical properties of the steels seems to show that AISI 316 L, AMCR (a Cr-Mn steel) and DIN 1.4914 are not subject to Liquid Metal Embrittlement (LME) in the presence of Pb-17Li at 523 or 573 K /35/, (the melting poing of the embrittler is 508 K).
Stress corrosion tests carried out on tensile specimens of AISI 316 L, in Pb-17Li for 2500 h at 723 K with constant loads corresponding to 80 and 60 % of the ultimate stress have not revealed any influence of stress on corrosion /35/. Equivalent results are reported /36/ after 6000 h at 673 K with constant uniaxial stress of 134 MPa.

2.4. Corrosion caused by interaction between the first wall/blanket structural material and solid lithium compounds.

There are limited data on the compatibility of the candidate solid breeding materials with potential structure or cladding material. The thermodynamic stability of the oxides considered: Li_2O, $LiAlO_2$, Li_2SiO_3, Li_4SiO_4, Li_2ZrO_3 indicate that they should not react directly with Fe-Cr alloys. The main problem concerns the impurities present in these materials mainly LiOH or Li_2CO_3 introduced during fabrication or resulting from oxygen and hydrogen partial pressures of the environmental atmosphere. Tests carried out in closed systems at 873-1000 K /37,38,39/ have revealed that Li_2O is more reactive with steels than $LiAlO_2$, Li_2SiO_3. The reaction products detected were $LiCrO_2$, Li_5FeO_4 and $LiFeCrO_4$. More recently /40/ it was shown that the reaction of Li_2O with AISI 316 in the temperature range 773-973 K could be avoided after scrupulous re-purification of Li_2O with Li to remove all traces of LiOH. Lately /41/, tests were carried out to investigate (always in a closed system) the compatibility of various Li based oxides with AISI 316 and the ferritic-martensitic steel 1.4914, in the temperature range 773-1073 K for times up to 500 h, in the presence of various amounts of H_2O or NiO as initial contaminants. It appears that, in addition to temperature and time, the chemical interaction between solid breeder materials and the Cr-Ni steels depend on the moisture content or oxygen partial pressure in the system.
The extent of cladding attack caused by NiO is comparable to that caused by H_2O; it is thus stated that "the formation of LiOH does not play the decisive role but the presence of reactive oxygen introduced by H_2O or NiO". When a sufficiently high oxygen potential is present, which results in the formation of Cr and Fe oxides or spinels the latter react with the Li compound. The attack of the cladding is stronger for Li_2O followed sequentially by Li_4SiO_4, Li_2SiO_3, $LiAlO_2$ and Li_2ZrO_3. Fig. 15 illustrates this trend. If a cladding attack of 200 μm after 10 000 h (H_2O content 1 mole % per 1 mole Li_2O) is acceptable, the following provisional maximum cladding temperatures are given for both AISI 316 and 1.4914 steels: Li_2SiO_3, 1173 K; Li_4SiO_4, 1063 K; Li_2O, 993 K. Tests in a flowing gas environment with controlled H_2O partial pressure still have to be performed. It should also be mentioned that sulphidisation of AISI 316 has been observed /26/ in contact with $LiAlO_2$ containing sulphate impurities.

REFERENCES

1. UCRL-50647.
2. HEDL-TME 78-15-UC 20, April 1978
3. ANL 8001, March 1973
4. ORNL/TM-4927, January 1976
5. E.E. Hoffman, ORNL-2924 (1960)
6. R.A. Patterson, R.S. Schlager, O.L.Olson, J. Nucl. Mat. 57 (1975), 312

7. D.L. Olson, G.M. Reser, D.M.Matlock, NACE, 35 (1980)

8. M.G. Barker, S.A. Franklin, P.G.Gadd, D.R. Moore, Material Behaviour & Physical Chemistry in Liq. Metal Systems, Ed. H.U. Borgstedt (Plenum N.Y. 1982) p.113

9. E. Ruedl, V. Coen, T.Sasaki, H.Kolbe, J. Nucl. Mat. 110 (1982), 28

10. E. Ruedl, T. Sasaki, J. Nucl. Mat. 116 (1983), 112

11. E. Ruedl, V. Coen, T.Sasaki, H.Kolbe, J. Nucl. Mat. 123, Nos1-3, May 1984, 1247

12. O.K.Chopra, D.L.Smith, to be published in J. Nucl. Mat.

13. J.H.Devan, J. Nucl. Mat. 85&86 (1979), 249

14. V.Coen, H.Kolbe, L.Orecchia, T.Sasaki, J. Nucl. Mat. 85&86 (1979), 271

15. P.F.Tortorelli, J.H.Devan, R.M.Yonco, J. Mat. for En. Syst. Vol.2, March 1985, 5

16. O.K.Chopra, D.L.Smith, J. of Nucl. Mat. 133&134 (1985), 861

17. P.F. Tortorelli, Personal Communication, work to be published in J. of Nucl. Mat.

18. ANL/FPP-84-1, Vol.2, ch.6

19. H.Tas, J.A. Dekeyser, F.Casteels, J. Walnier, F.Deschutter, FT/Mol/T86-04

20. O.K.Chopra, J. Nucl. Mat. 115 (1983), 223

21. M.Grundmann, H.U. Borgstedt, KFK 3819, Oct. 1984

22. P.Fenici, V.Coen, J.Arrighi, H.Kolbe, T. Sasaki, E.Ruedl, J. of Nucl. Mat. 85&86 (1979), 277

23. P.Fenici, V.Coen, E.Ruedl, H.Kolbe, T. Sasaki, J. of Nucl. Mat. 103&104 (1981), 689

24. V.Coen, J. Nucl. Mat. 133&134 (1985), 46

25. V.Coen, P.Fenici, H.Kolbe, L.Orecchia, T. Sasaki, J. Nucl. Mat. 110 nº1, Sept. 1982, 108

26. P.Fauvet, J.Sannier, G.Santarini, Proceedings of the 13th SOFT, 1984, Pergamon Press

27. H.U.Borgstedt, G.Frees, G. Drechsler, to be published in J. Nucl. Mat.

28. M.Broc, P.Fauvet, T.Flament, J.Sannier, to be published in J. Nucl. Mat.

29. O.K.Chopra, D.L.Smith, J. Nucl. Mat. 122 & 123 (1984), 1219

30. F.Tortorelli, J.H.Devan, to be published in J. Nucl. Mat.

31. O.K.Chopra, D.L.Smith, to be published in J. of Nucl. Mat.

32. V.Coen, A.T. Dadd, H.Kolbe, L.Orecchia, Proc. 3rd Int. Conf. on Liq. Met. Eng. and Technology, Oxford (April 1984) The British Nucl. Soc. London, Vol.1 p.347

33. W.R.Watson, R.J.Pulham, ibid. Vol. 3, p. 99

34. M.G. Barker, V. Coen, J.A. Lees, H. Kolbe, L. Orecchia and T. Sample, to be published in J. of Nucl. Mat.

35. V.Coen, H.Kolbe, L.Orecchia, to be published in J. of Nucl. Mat.

36. M.Broc, T.Flament, P.Fauvet, J.Saunier, Personal Communication, work to be published in J. of Nucl. Mat.

37. T.Kurasawa, H.Takeshita, S.Muraoka, S.Nasu, M.Miyake, T.Sano, J. Nucl. Mat. 80 (1979), 48

38. P.A.Finn, S.R.Breon, N.R.Chellew, J. Nucl. Mat. 103&104 (1981), 561

39. O.K.Chopra, D.L.Smith, J. Nucl. Mat. 103 & 104 (1981), 555

40. R.J.Pulham, W.R.Watson, J.J.Collinson, J. Nucl. Mat. 122&123 (1984), 1243

41. P.Hofmann, W.Dienst, to be published in J. Nucl. Mat.

SCHEMATIC DIAGRAM OF TYPICAL NUCLEAR FUSION REACTOR CYCLES

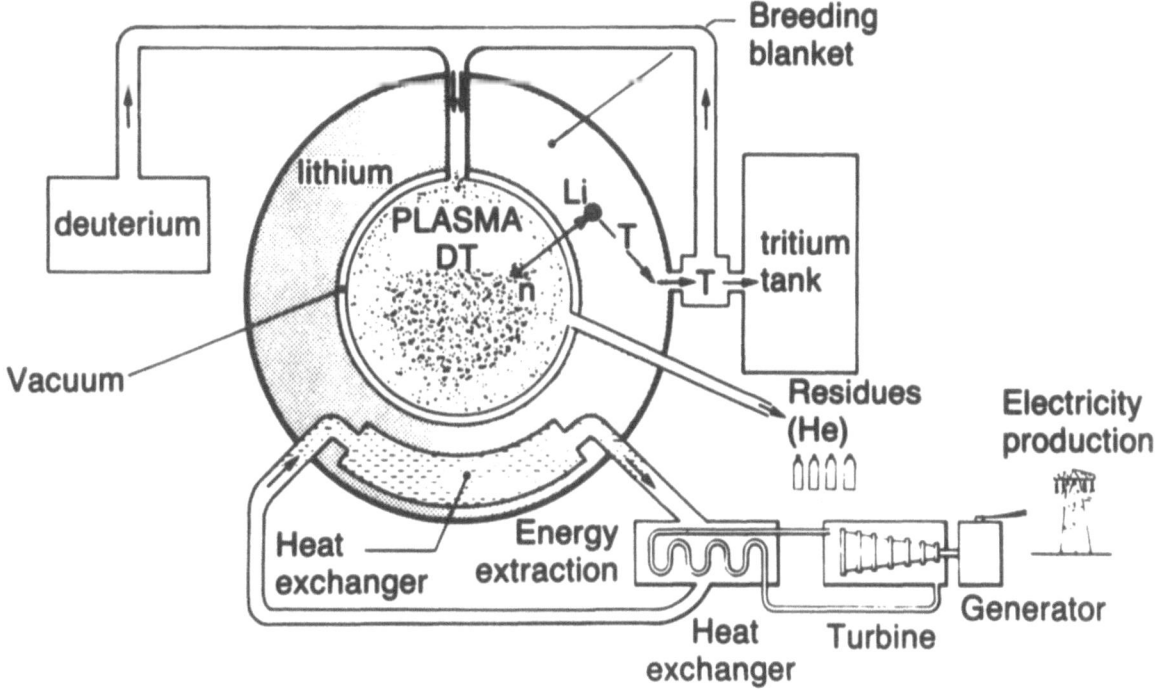

Fig. 1 Schematic diagram of typical nuclear fusion reactor cycles.

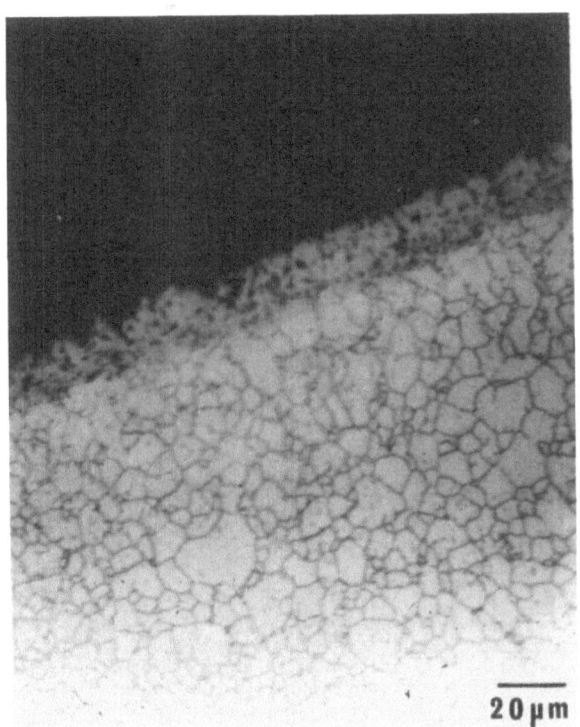

Fig. 2 Optical micrograph of cross section of a sample of AISI 316 exposed to Li (with fairly high N level) for 90 days at 873 K.

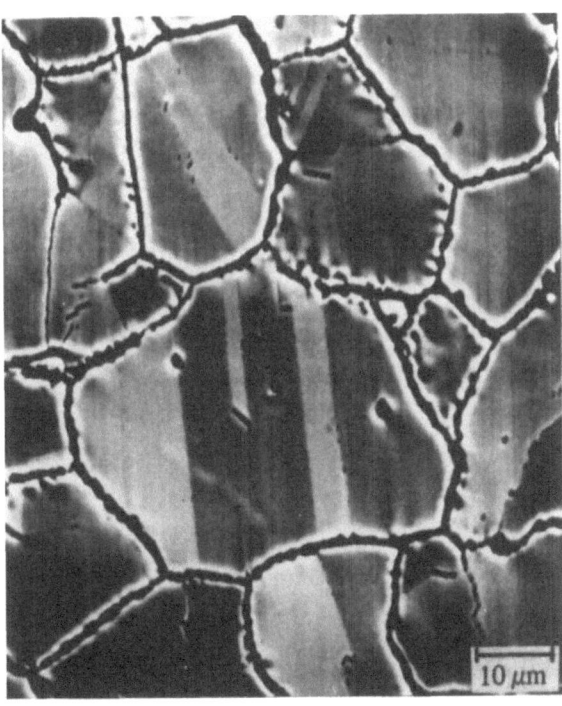

Fig. 3 SEM micrograph of cross-section of Nitronic 32, heat treated in Li at 873 K for 1000 h. BEI composition.

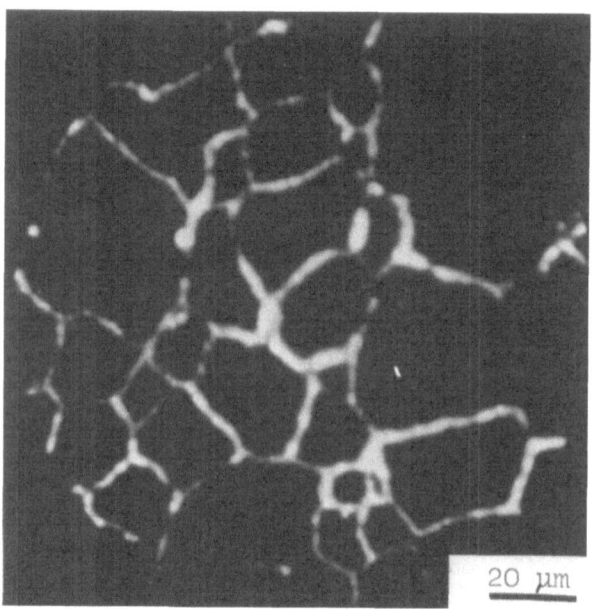

Fig. 4 Cross section of Carpenter 18/18 Plus heat treated in Li at 873 K for 1500 h. Secondary $^7Li^+$ ion image as obtained in ion probe microanalyser. Primary ions $^{32}O_2^+$.

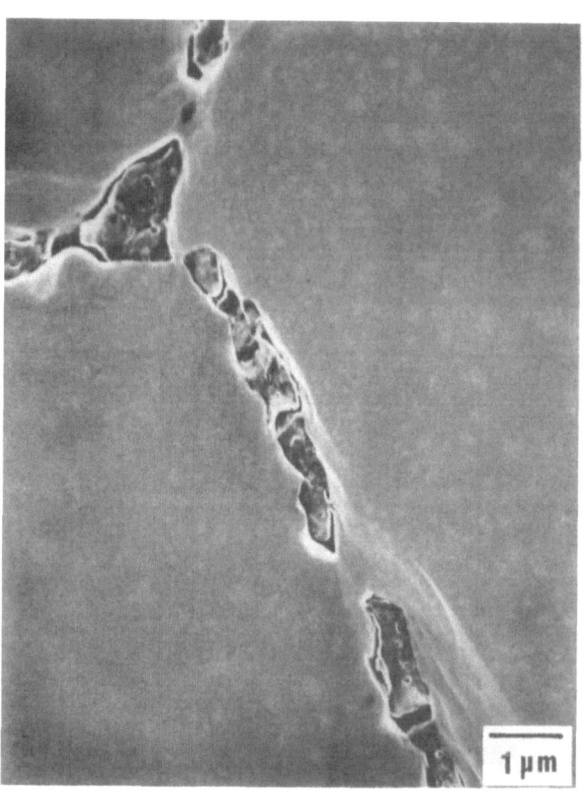

Fig. 5 SEM micrograph of cross-section of Nitronic 32 heat treated at 873 K in Li for 1000 h. Note formation at the grain boundaries of the Li-Cr-N compound.

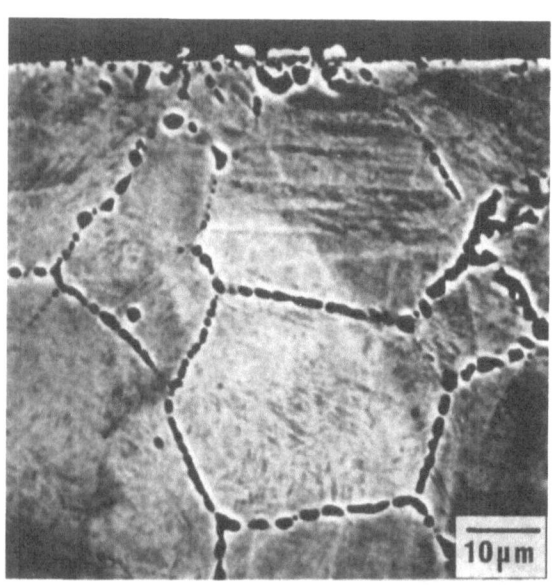

Fig. 6 SEM micrograph of cross-section of Nitronic 32 heat treated in Li at 873 K for 1500 h. BEI composition.

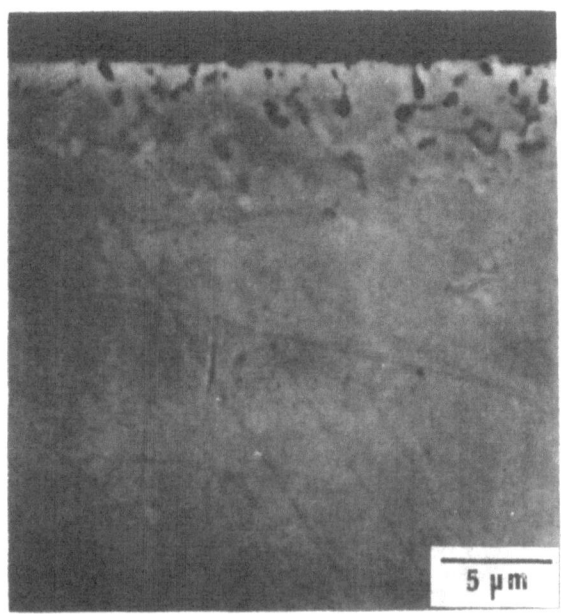

Fig. 7 SEM micrograph of cross-section of Nitronic 32 heat treated in Li + 40 ppm H in Li at 873 K for 1500 h. BEI composition.

Li / AUSTENITIC S.S.

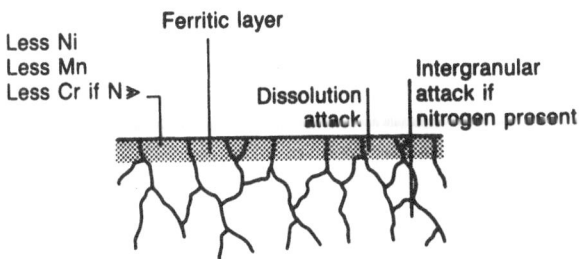

Formation of Li_9CrN_5 on grain boundary after a delay time varying from alloy to alloy

Presence also of $M_{23}C_6$ carbides **richer in Cr** and **poorer in Mn and Fe**

No oxygen effect

Fig. 8 Corrosion mechanisms for the Li/austenitic stainless steel couple.

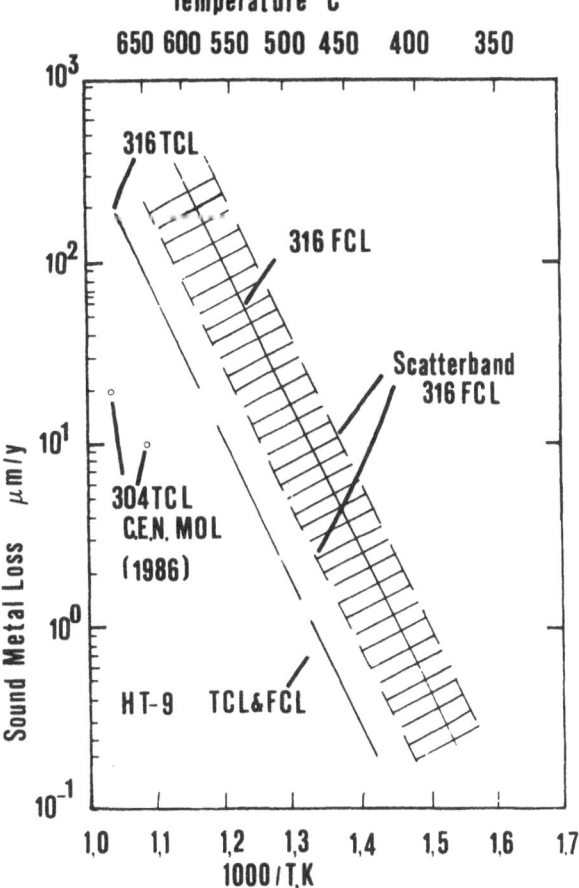

Fig. 9 Arrhenius plot of corrosion rate data for type 304 and 316 stainless steel and HT-9 taken from Ref. 18 plus CEN Mol results.

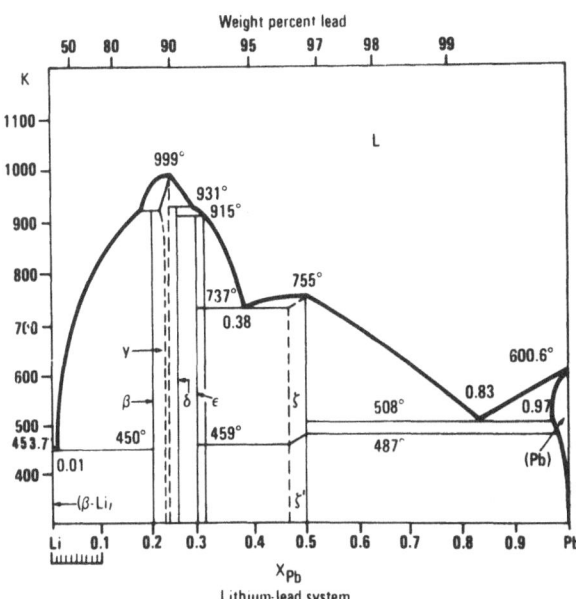

Fig. 10 Lithium-lead system.

43

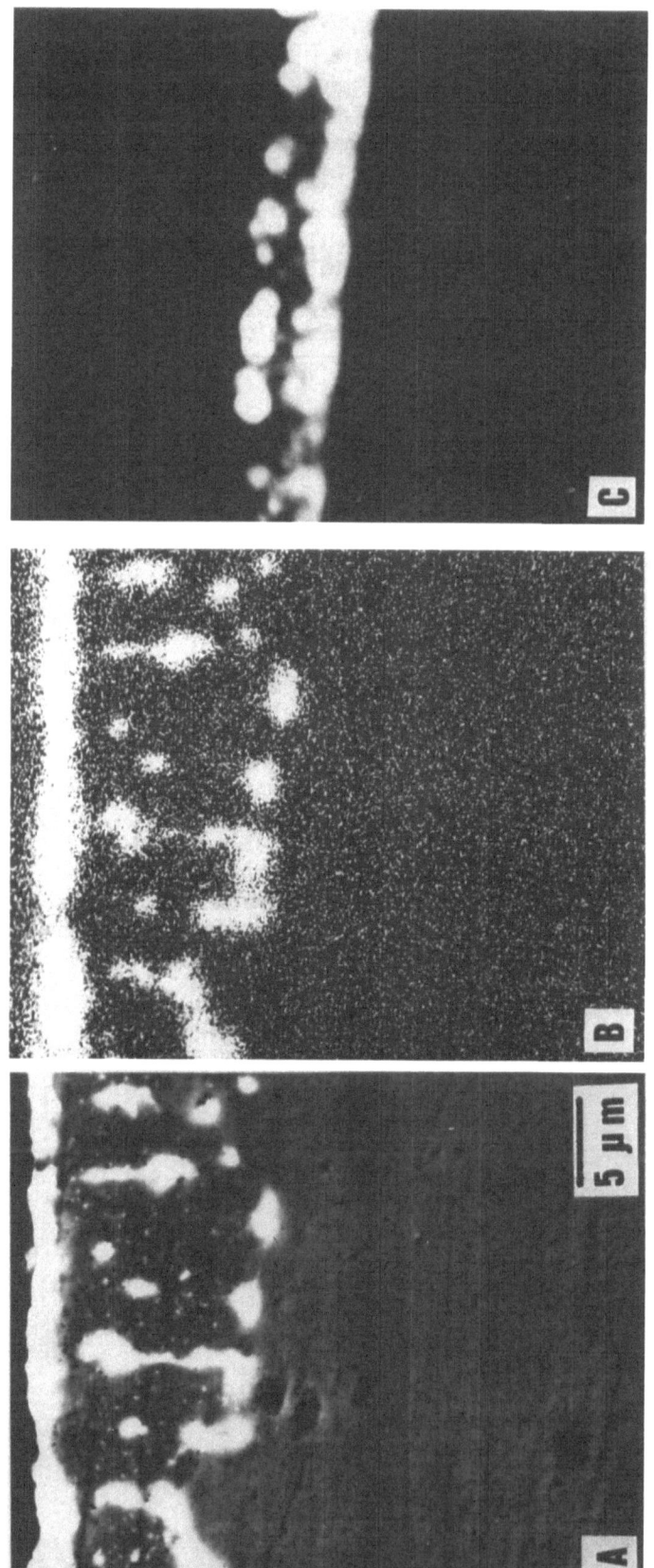

Fig. 11 SEM micrograph of crosssection of
sample of AISI 316 L heat treated in
Pb-17Li at 873 K for 750 h showing Pb
and Li penetration.

a) BEI composition

b) Pb-Mα X-ray image

c) Secondary ion image of $^7Li^+$. Primary
   ions $^{32}O_2^+$;diameter of analyzed zone
   = 120 μm.

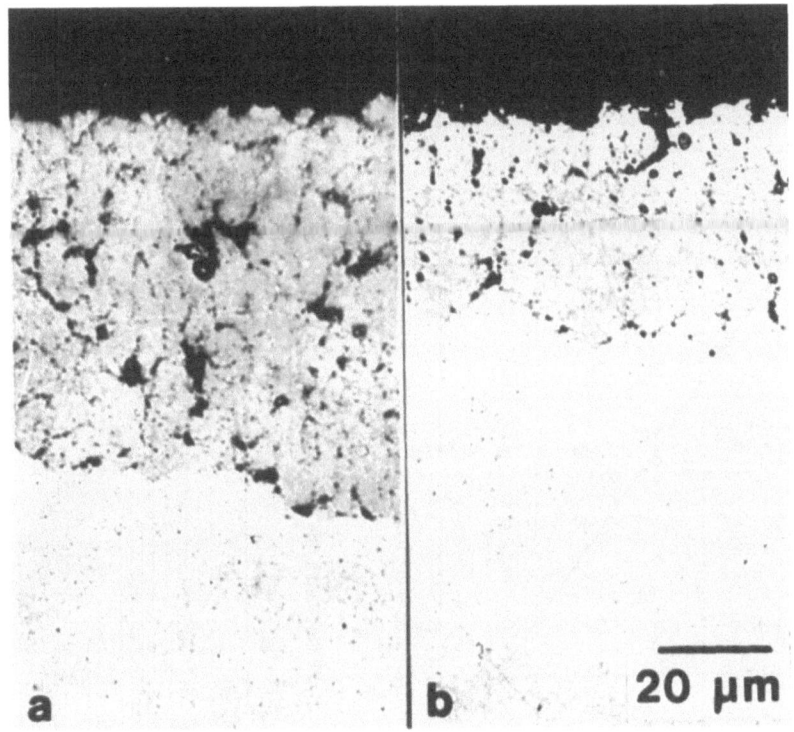

Fig. 12 Optical micrographs of cross section of
316 L steel exposed to Pb-17Li for
320 h at 873 K
a) with added oxygen, b) without added
oxygen.

Pb-17Li / AUSTENITIC S.S.

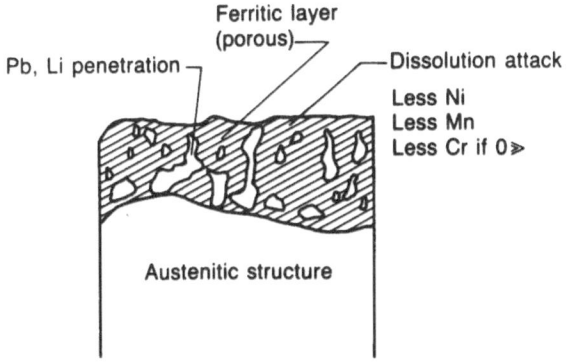

No grain boundary penetration
No nitrogen effect
Increase of dissolution rate in presence of O
Formation of a surface Li-Me-O ?
              LiCrO₂ ?

Fig. 13 Corrosion mechanisms for the Pb17Li/-
austenitic stainless steel couple.

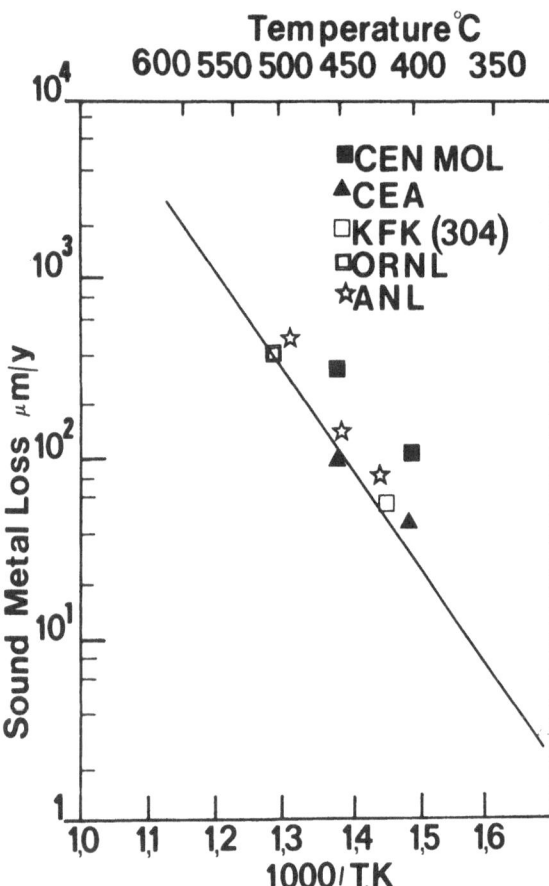

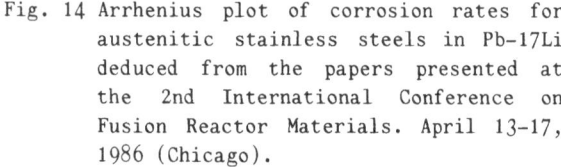

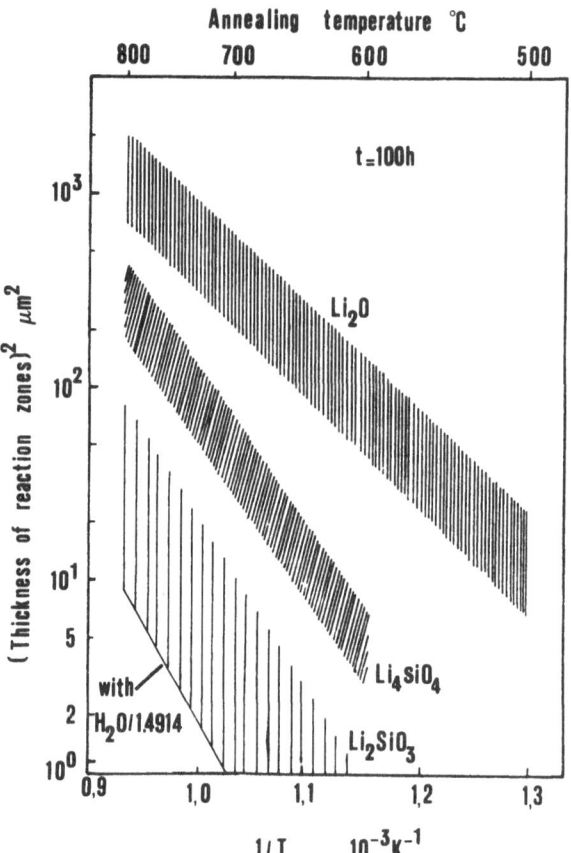

Fig. 14 Arrhenius plot of corrosion rates for austenitic stainless steels in Pb–17Li deduced from the papers presented at the 2nd International Conference on Fusion Reactor Materials. April 13–17, 1986 (Chicago).

Fig. 15 Chemical reactions of Li-based oxide ceramics containing 1 mol%H_2O or 1 mol%NiO per 1 mol Li_2O with various stainless steels (11–17%Cr) normalized to an annealing time of 100 hours, from Ref. 41.

8 Corrosion Resistance of Metallic Materials for Use in Nuclear Fuel Reprocessing

J. Pinard Legry, M. Pelras and G. Turluer

The authors are in the Corrosion Department os the COMMISSARIAT A L'ENERGIE ATOMIQUE at Fontenay-aux-Roses (France). G. PINARD LEGRY is manager of this Group with M. PELRAS and G. TURLUER as expert engineers for corrosion in reprocessing plants. G. TURLUER recently moved to the Safety Analysis Department.

SYNOPSIS

This paper reviews the corrosion resistance properties required from metallic materials to be used in the various developments of the PUREX process for nuclear fuel reprocessing. Stainless steels, zirconium or titanium base alloys are considered for the various plant components, where nitric acid is the main electrolyte with differing acid and nitrate concentrations, temperature and oxidizing species.

1. INTRODUCTION

Corrosion resistance properties required for materials to be used in the various developments of the Purex process are crucial in the field of nuclear field reprocessing, since equipments inspection, decontamination, repair or premature replacement would result in procedures much more complicated than in conventional chemical industries (1).

This paper will review the various stages of reprocessing common to PWR, BWR, Magnox, breeder elements, where corrosion may be a problem, the principal materials compatible with the various nitric solutions used in full implementation of the Purex process, their limitations in corrosion resistance, and possible ways to mitigate or avoid them.

2. PROCESS DESCRIPTION: IDENTIFICATION OF CORROSION SENSITIVE COMPONENTS

As shown schematically in Fig. 1, irradiated fuel elements containing uranium, plutonium and fission products are dissolved in hot or boiling nitric acid, leaving behind chopped hulls and insoluble dissolution fines: Pu and U are further separated in the course of one or several extractions processes at low temperatures $\leqslant 50°C$ between organic solvents containing tributylphosphates (TBP) and nitric acid aqueous solutions of low or moderate acidity.

The corrosion sensitive operations with regard to stainless steels are to be found almost exclusively in equipments for heating and containing hot or boiling concentrated nitric solutions or nitric solutions where highly oxidizing species such as Fe^{3+}, Pu (VI) and possibly Cr (VI) are present at a critical concentration.

Such conditions exist in the following types of equipment (represented in dashed lines in Fig. 1):

. acid dissolver (U + Pu, 3 - 13 N HNO_3)
. acid recovery evaporators (8 - 13 N HNO_3)
. intercycle U + Pu evaporator [2 - 3 N HNO_3), U, Pu (VI)]
. final plutonium nitrate concentrator [5 N HNO_3 + Pu $\rightarrow$ Pu (VI)]

- oxalic mother liquor evaporator (HNO_3 up to 14 N)
- vitrification off gas treatment (6 N HNO_3 + Fe (III) + Cr)

In comparison with Magnox fuels, the reprocessing of PWR, BWR and in particular fast breeder fuels with much higher burn-ups, yields large amounts of dissolution fines made up of platinoids such as platinum, ruthenium and rhodium; these elements may enhance stainless steel corrosion possibly in the transpassive state by local galvanic coupling in concentrated nitric acid on hot spot areas.

Furthermore, fast breeder fuel dissolution produces higher plutonium contents and also by partial corrosion of the stainless steel hulls corrosion products such as Fe^{3+}, Cr^{3+}: these ions may then enhance the corrosivity of the dissolution liquors particularly in the final stages of a batch process.

3. RELEVANT MATERIALS: EVALUATION OF THEIR CORROSION RESISTANCE

Adequate corrosion resistance in the large range of subazeotropic nitric acid solutions of variable redox properties is one prerequisite not only for the material but also for a finished component. Industrial availability and know-how are also essential to ensure reliably all stages from elaboration to the various transformations, shaping, assembly, welding. These conditions are met for the more widely used material class i.e., unstabilized low carbon stainless steels, and are just being developed for titanium and zirconium in the chemical industry. Typical compositions of those materials and of others still at the development stage are given in Table I.

Illustration of the comparative corrosion resistance of some alloys in four nitric acid solutions of differing oxidizing power is given in Table II which reports short duration CEA test results.

3.1 Low carbon austenitic steels

Since corrosion resistance is essentially determined by the chromium content, nickel base and molybdenum additions being of little interest in nitric acid, the most widely used are unstabilized low carbon steels of the type AISI 304 L (AFNOR Z 2 CN 18-10) or AFNOR Z 2 CN 25-20 akin to an extra low carbon (ELC) type AISI 310 (2-15).

The use of stabilized type AISI 347 with niobium) and AISI 321 (with titanium) has been reconsidered by BNFL in favour of a special nitric acid grade (NAG) of the type AISI 304 L, since knife line attack has been found on welded structures (3).

Even very low carbon unstabilized stainless steels may undergo an intergranular attack (IGA) in two different types of conditions:

- When sensitized by a thermal treatment or welding if carbon is locally enriched as a result of a uneven distribution or

surface contamination in the transformation stages.

When the medium becomes oxidizing enough so as to bring the material into the transition between the passive and trans-passive states or even right into the transpassive range, in which accelerated general corrosion also occurs.

In concentrated boiling nitric acid, IGA occurring even on fully annealed material is all the more deleterious since the corrosion products such as Fe and Cr (susceptible to oxidation to Cr^{6+}) can build up and exert a self-accelerating effect on the attack.

In fuel reprocessing, IGA is even more likely to occur as acid concentration, process temperature and wall temperature (increased by heat flux) are higher and oxidizing ions such as Cr (VI), Pu (VI) and higher concentrations of Fe can be produced.

Advent of low impurity grades

New grades have been developed for nitric acid applications to reduce the content in minor elements such as C, Si, P, S, which by preferential segregation in the grain boundaries decrease the corrosion resistance with respect to IGA (5, 6, 10, 11, 15).

Implementation of the necessary refining treatment by argon vacuum remelting (ASV) (4, 5, 7) or electroslag remelting (ESU) (10, 11) also improves inclusion cleanliness, which ensures minimal hazards with respect to end grain attack.

Work in progress on improved materials such as Uranus 16 or Werkstoff n° 1.4306 S (ESU) concerning the industrial feasibility and reliability and laboratory qualifications should be extended to demonstrate that the achieved concentrations of residuals and of silicon in particular, are sufficiently low to minimize, if not suppress, the occurrence of IGA.

High silicon stainless steels: Cr Ni Si 17-15-4

When immunity to intergranular attack and therefore corrosion resistance predictability in strongly oxidizing media is required and a somewhat relatively higher general corrosion rate can be tolerated provided by the relevant corrosion allowance, the use of high purity grades containing at least 4% silicon such as Uranus S1N (Z 1 CNS 17-15) offers a safer choice (4, 7, 11, 16, 17) as shown by corrosion data in Table II.

The effect of silicon additions on the corrosion resistance of non-sensitized high purity 14% Cr - 14% Ni alloys in HNO_3 containing Cr^{6+} ions has been summarized by Armijo and Wilde in Fig. 2.

3.2 Titanium and zirconium base alloys

Titanium finds its best corrosion resistance properly when the nitric acid media become too corrosive for the improved stainless steel grades susceptible to transpassive corrosion, or when silicon rich steels would display prohibitive general corrosion rate (11, 18, 20), i.e. in the strongly oxidizing and acid media as shown in Table II.

However titanium displays a moderate general corrosion resistance in less oxidizing boiling acids, in renewed liquid phases and when submitted to high trickling rates of condensate; a maximum corrosion rate around 80 mdd (0.640 μm.year^{-1}) is observed for boiling 11 N HNO$_3$.

Alloying with tantulum as with Ti 5% Ta has been proposed to improve in a relative manner the weak general corrosion resistance in renewed media (19).

Zirconium alloys: grades 702 and 704

Even improved resistance of Ti 5% Ta cannot compete with the extremely low corrosion rates usually $<$ 1 mdd throughout the whole range of subazeotropic concentrations, redox and temperatures prevailing in fuel reprocessing (11, 19, 20, 22, 23), experienced with grades 702 and 704.

The occurrence of stress corrosion cracking in these grades in 90% HNO in severely strained C-rings or U-bends (11, 21, 22) is not observed in this stressing procedure in the subazeotropic compositions relevant to fuel reprocessing (20, 33, 23).

Both zirconium and titanium alloys are more sensitive to the presence of uncomplexed fluoride than would be stainless steel; this element only causes a general corrosion enhancement of zirconium at concentrations exceeding 1 ppm.

4. WAYS TO MITIGATE OR AVOID CORROSION PROBLEMS

Corrosion problems can be mitigated or avoided through different approaches which include the following measures:

- a diversified material selection specific to each sensitive component as exemplified in France by COGEMA and SGN which in specific cases select Z 2 CN 18-10, Z 2 CND 17-13 (8), Z 1 CNNb 25-20 (as Uranus 65), Z1 CNS 17-15 (as Uranus S1N) stainless steels and now zirconium essentially for the most highly oxidizing high temperature concentrated acid solutions (4, 7, 8, 20);

- a modification of process variables to minimize the corrosivity such as:

  . operation of acid recovery under reduced pressure
  . limitation of heat flux to reduce wall temperature
  . operation of an intercycle concentration after plutonium removal
  . cooling and stirring, to avoid hot spot corrosion under deposits (8)
  . discharge of concentrates as a function of oxidizing ions build up.

- design improvements to avoid crevice effects, unrenewed volumes, unfavorable condensate effects and location of sensitive junctions (for example zirconium/stainless steels) in high temperature processes;

- improved fabrication and assembly procedures so as, for example, to minimize tunnel corrosion and avoid defective welding procedures.

5. CONCLUSION

The lessons gained from laboratory and industrial experience, the implementation of improved corrosion resistant materials, together with more stringent acceptance tests make it now possible to restrict if not eliminate the recurrence of some premature corrosion damages observed in some first generation plants.

Some scope of improvement still remains with respect to the industrial elaboration and transformation of very low impurity austenitic stainless steels, to an extended implementation of more noble materials and to development of new acceptance tests more closely related to service applications.

REFERENCES

(1) OTANI, YOSIKUNI, Donen Giho, 53, 63 (1985)
(2) M. PELRAS, Bulletin d'Informations Scientifiques et Techniques du CEA, n° 139, 31 (1969)
(3) R.D. SHAW and D. ELLIOTT, Stainless steel'84 Conference, Goeteborg (Sweden) (1984)
(4) H. CHAUVE, J. DECOURS, R. DEMAY, M. PELRAS, J. SIMONET, Nuclear Europe, 2, 19 (1986)
(5) A. DESESTRET, J. FERRIOL, G. WALLIER, Matériaux et Techniques, Sept/Oct. (1977)
(6) A. DESESTRET, G. GAY and P. SOULIGNAC, 25ème Colloque de Métallurgie de Saclay, June (1982)
(7) J. DECOURS, J.C. DECUGIS, R. DEMAY, M. PELRAS, G. TURLUER, Paper TC 590-8, IAEA Technical Committee Meeting on materials reliability in the back end of the nuclear fuel cycle, Vienna (Austria), Sept. (1986)
(8) J. BACHELAY, J. DECOURS, R. DEMAY, M. PELRAS, L. ROZAND, G. TURLUER, Paper TC-580-11, IAEA Technical Committee Meeting on materials reliability in the back end of the nuclear fuel cycle, Vienna (Austria), Sept. (1986)
(9) S. LEISTIKOW, R. KRAFT, R. SIMON, KFK Report n° 3 740 (1984)
(10) R. SIMON, S. SCHNEIDER, S. LEISTIKOW, KFK Nachrichten, 2, 90 (1986)
(11) E.M. HORN and H. KOHL, Werkst.u.Korrosion, 37, 57 (1986)
(12) U. BLOM and G. KVARNBACK, Materials Performance, 43, July (1975)
(13) B.E. PAIGE, Materials Performance, 12, 22 (1976)
(14) N.W. WILDING and B.E. PAIGE, Report ICP 1 107, Allied Chemical Corporation, ERDA, Dec. (1976)
(15) O.V. KASPAROVA, S.D. BOGOLYUBSKII, Ya. KOLOTYRKIN, V.M. MIL'MAN, Zashita Metallov, 20, 6, 844 (1984)

(16) A. KRATZER, B. PIEGER, H. TISCHNER, E.M. HORN, Proceedings 9th ICMC, 2, 465, Toronto (USA), June (1984)

(17) R.R. KIRCHHEINER, F. HOFMAN, Th. HOFFMANN, G. RUDOLPH, Corrosion 86, Paper 120, NACE, Houston (USA) (1986)

(18) TAKAMURA, K. ARKAWA, Y. MORIGUSHI, The Science Technology and Application of titanium, R. Jaffee Editor, Pergamon Press, Oxford (Great Britain) (1970)

(19) F. FURUYA, H. SATO, K. SHIMOGORI and al., Proceedings ANS Meeting on Fuel Reprocessing and Waste Management, Jackson Wy. (USA) (1984)

(20) H. CHAUVE, J. DECOURS, R. DEMAY, M. PELRAS, J. SIMONNET, G. TURLUER, Paper TC 580-9, AIEA Technical Committee Meeting on materials reliability in the back end of the nuclear fuel cycle, Vienna (Austria), Sept. (1986)

(21) J.A. BEAVERS, J.C. GRIESS, W.K. BOYD, Corrosion, 36, 5, 292 (1981)

(22) T.L. YAU, Corrosion, 39, 5, 168 (1983)

(23) T.L. YAU, 4th ASTM Symposium on Industrial Applications of Titanium and Zirconium, Philadelphia, Pa. (USA) (1984)

TABLE I

Materials and typical composition in weight %

MATERIALS	Cr	Ni	Mo	Mn	C	Si	P	S	N	Nb
304 L ≃ Z 2 CN 18-10 ≃ W n° 1.4306 n	17.5/19	9/12	<0.2	0.5/2	<0.025	0.2/1	<0.045	<0.03		
W n° 1.4306 s (réf. 10)	19	12.5	0.02	1.7	0.015	0.02	0.02	0.007		
W n° 1.4306 s (ESU) (Réf. 10)	19	12.5		1.6	0.007	0.02	0.02	0.007		
NAG 18-10 L (réf. 3)	17.5/19	9/11	<0.2	1/2	<0.025	0.2/0.8	<0.02	<0.015		
Uranus 16 (ASV) (réf. 6) (Z 2 CN 18-10)	18	10			<0.015	<0.15	<0.015	<0.005		
Uranus 65 (ASV) (réf. 7) (Z 1 CN 25-20)	24/26	19/22	<0.5	<2	<0.015	<0.20	<0.02	<0.005		addition
Uranus S1N (ASV) (réf. 7) (Z 1 CNS 17-15)	16.5/18.5	13.5/15	0.5	<2	<0.015	3.8/4.5	<0.02	<0.005	<0.035	addition

	Zr+Hf min.	Hf max.	Fe+Cr	Sn	Nb	Cu max.	C max.	H max.	N max.	O max.
Zirconium grade 702	99.2	5	0.2			0.01	0.05	0.005	0.01	0.18

	Ti	Ta	Fe			H	N	O
Titanium grade 2 (réf. 19)	bal.		0.05			0.0025	0.003	0.09
Ti- 5 % Ta (réf. 19)	bal.	4.75	0.05			0.0025	0.005	0.09

TABLE II

Comparison of typical mean general corrosion rates
in 4 boiling nitric acid solutions of increasing oxidizing power ;
weight loss in $mg.dm^{-2}.day^{-1}$ (mdd)
[(IGA) when observed = intergranular attack]

MATERIALS	PERIODICALLY RENEWED BOILING MEDIA : 5 x 48 HOURS							
	5 N HNO_3		14.4 N HNO_3		5 N HNO_3 + 1 Ml^{-1} Pu		5 N HNO_3 + 1 g.l^{-1} Cr	
	mdd	$\mu m.year^{-1}$	mdd	$\mu m.year^{-1}$	mdd	$\mu m.year^{-1}$	mdd	$\mu m.year^{-1}$
Quench-annealed stainless steels								
Z 2 CN 18-10 (AISI 304 L)	2	10	30	135			780 (80 h.)	3 510 (IGA)
Z 2 CNNb 25-20 (as Uranus 65) akin to 310 ELC	1	5	20	90	180 (96 h.)	810 (IGA)	1 100 * (80 h.)	4 950 (IGA)
Z 1 CNS 17-15 (as Uranus S1N)	20	90	130	590			30	135
Titanium	30	240	20	160	< 2	< 16	< 2	< 16
Zirconium grade 702	<0.5	<3	<0.5	<3	< 2	< 10	< 0.5	< 3

* actually 8 N HNO_3 + 1 g.l^{-1} Cr (VI)

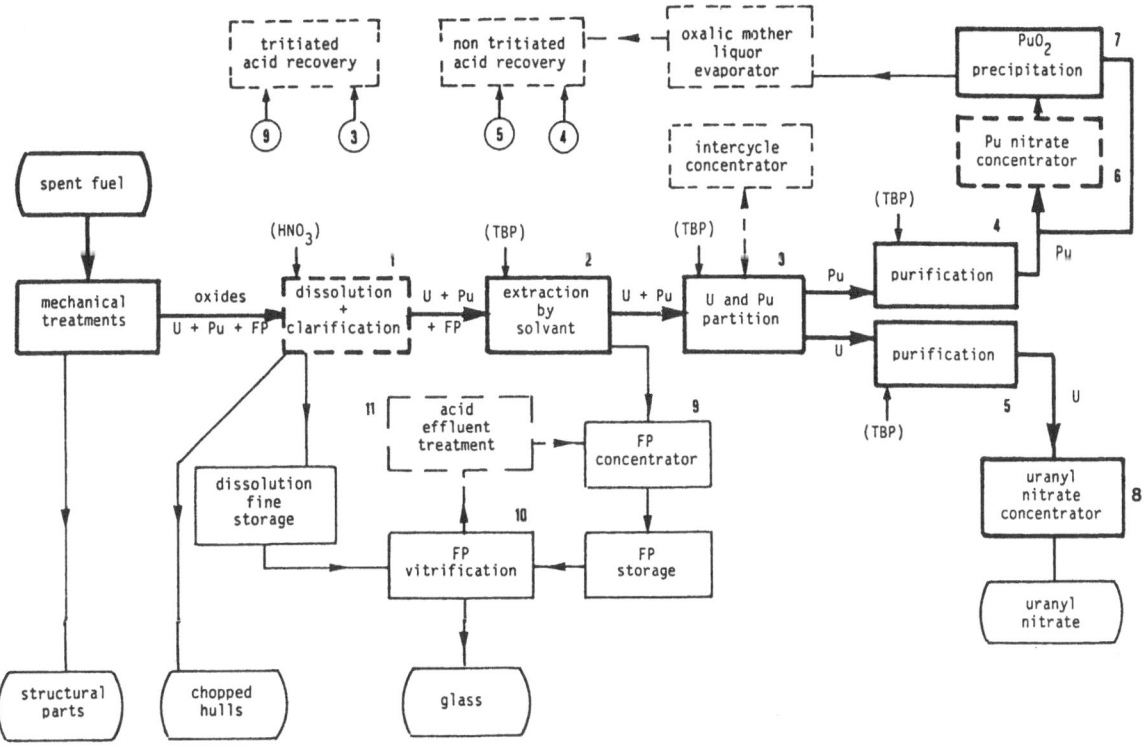

1 Schematic diagram of a Purex based fuel reprocessing featuring some corrosion sensitive parts of the process and of fluid recovery.

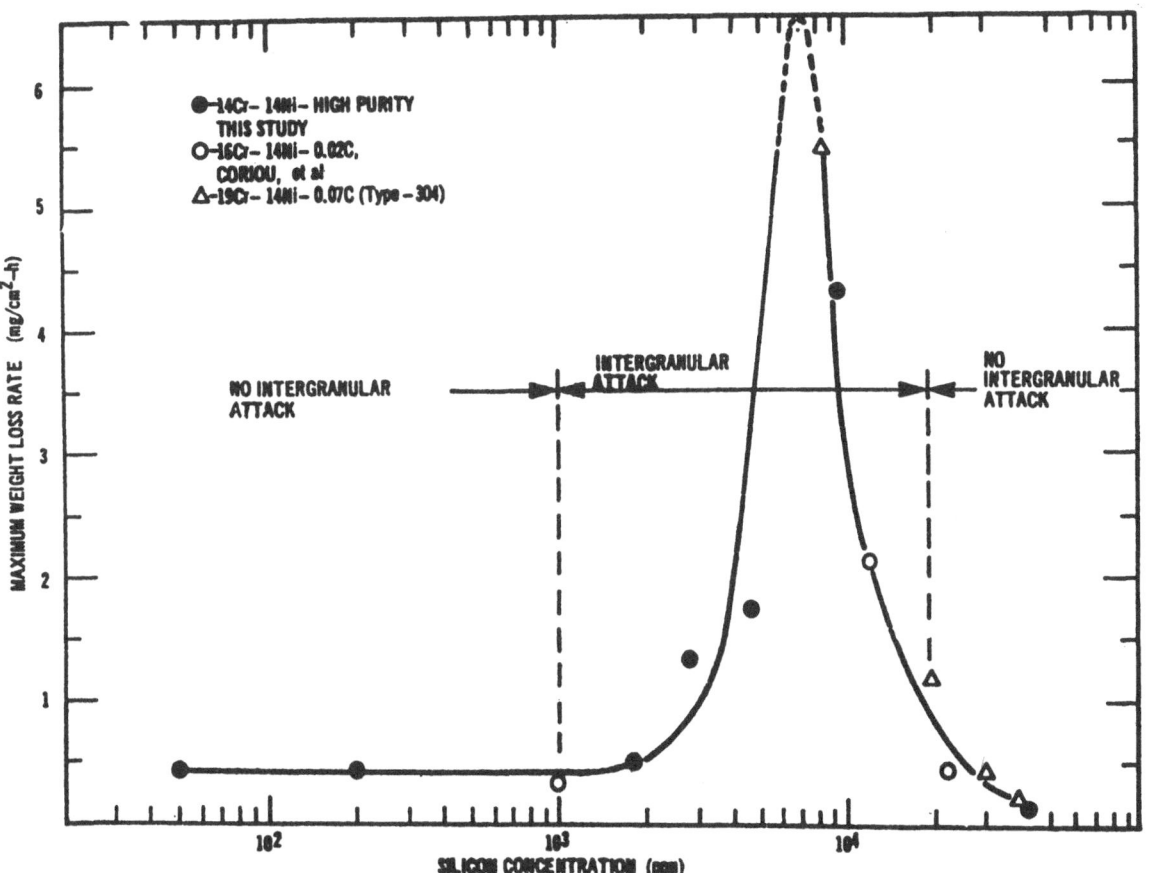

2 Effect of silicon additions on the corrosion resistance of non-sensitized high purity 14% Cr, 14% Ni, balance Fe alloys in HNO + Cr (From Armijo and Wilde, Corros. Sci.8,649-664 (1968)

9 Corrosion Aspects of Containers for High-Level Waste Disposal

J. Weber and J. P. Simpson

Sulzer Brothers Ltd.,
Winterthur, Switzerland.

SYNOPSIS

Nuclear power produces a range of waste types which require safe disposal. Most countries have set themselves the task of disposing of nuclear waste in repositories on their own soil.

The aim of all projects is to prevent radionuclides from entering the biosphere. Most concepts for the disposal of high level waste (HLW) require that the technical barriers should be effective for from 10^3 to 10^6 years. This requirement could be fulfilled by a container. Corrosion specialists must forecast the behaviour of the container materials over these time periods.

INTRODUCTION

Atomic energy is an important source of electric power in many industrialised nations; the operation of a nuclear power plant produces a range of nuclear waste types which require safe disposal. Most countries have set themselves the task of disposing of the nuclear waste produced in repositories on their own soil. There is considerable international cooperation on points of common interest.

The aim of all projects is to prevent radionuclides from entering the biosphere. Most concepts for the disposal of high level waste (HLW) require that the technical barriers should be effective for from 10^3 to 10^6 years. This requirement could be fulfilled by a container. Corrosion specialists must forecast the behaviour of the container materials over these time periods; a task which is surely unique in the history of corrosion science.

A relatively small amount of HLW is produced per unit of nuclear generating power. A 1000 MW(e) power plant produces around 4 m³ (10 tonnes), conditioned (vitrified) HLW per year. The HLWs are mixtures of short- and long-lived radionuclides. The radiotoxicity initially decays rapidly - more than 99% reduction in the first 1000 years - and then more slowly at longer periods, before finally reaching levels similar to those found in naturally occurring minerals.

The stages typical for most waste disposal projects are shown in table 1. All final disposal concepts for HLW include a series of complimentary engineered and natural barriers. The barriers going out from the waste are:
1. The waste form, e.g. the glass matrix of vitrified waste:
2. The container:
3. The backfill, i.e. the material used to fill the space around the container. (Not present in all concepts.)
4. The geological formation in which the repository lies.
5. The remaining path from the chosen geological formation to the biosphere; this includes the surrounding geological formations, the hydrogeological situation i.e. the rate of exchange of groundwater with water in the biosphere.

The container is important in that it is a high integrity barrier, most of the other barriers are retardation and/or dilution barriers. The radionuclides can not migrate as long as the container remains intact. The integrity of the container should be retained during the operational phase of the repository, and preferably until well after the perturbations of conditions in the geological formation due to the working of the repository and the thermal energy from the waste have dissipated. The thermal phase lasts about 1000 years.

The surface temperature of the container depends on the concept and can range up to 250°C after the repository is sealed.

Most of the first choice container types in the disposal concepts are metallic, although in several countries the alternative of a ceramic container is still being pursued.

The metal containers can be divided into two types, the 'consumable' type where the lifetime is attained by providing sufficient wall thickness (e.g. carbon steel) and the 'non-consumable' type where only a few millimeters of corrosion resistant material are necessary (e.g. passive materials or thermodynamically stable metals). The corrosion resistant layer is applied externally to a pressure vessel; the underlying material is credited with no corrosion resistance.

NATIONAL SOLUTIONS

Table 2 gives a summary of the geological formations chosen in selected national programmes, the materials under closest consideration, and the lifetime requirement for the container.

There are three basic types of repository which give three fundamentally different corrosion media, these are:
1. crystalline rock formations,
2. dry salt formations,
3. clay formations.

Repositories in dry rock formations above the water table have been proposed, thereby avoiding corrosion and transport of radionuclides by ground waters.

Repositories in crystalline formations.

Most crystalline rock formations under consideration are saturated with groundwater. There is a suggestion to site a repository in granite above the normal groundwater level (UK) and the tuff repository (Table 2) is also of the 'dry' type. The waters expected in these hard rock repositories are similar in type but vary greatly in salinity, all are saturated in silicate. Sea water is often used as a high salinity comparison medium. Some sample analyses of waters featured in the various programmes are given in Table 3.

The corrosion behaviour of some of the candidate materials in crystalline groundwaters and sea water is summarized in table 4.

As can be seen in Table 4, there is considerable interest in carbon steels, (and cast irons) cast or wrought for HLW containers

for lifetimes of up to ca. 1000 years. The life is attained by providing sufficient corrosion allowance. The Swiss container design for instance has a 250 mm wall thickness, 50 mm of which is the corrosion allowance, the remaining wall thickness is required to sustain the pressures at depths of 1200-1500 metres. The experimental evidence is that the average corrosion rate in a range of groundwater salinities is under 20 µm/yr in the low oxygen conditions expected in deep repositories [1]. The main problem is whether this attack is uniform. Pitting rates producing forecast penetrations of 200 mm in 1000 years exposure have been observed in carbonate/chloride solutions which favour pitting [2,3]. The high initial pitting rates necessary for these high penetration depths have not been observed in anaerobic natural or synthetic groundwaters [1,3,4].

Titanium alloys grade 12 and Ti-0.2% Pd are the most favoured of the high corrosion resistant materials. These alloys are less sensitive to pitting than the Ni-Cr types and are also more resistant to pitting at higher temperatures than pure titanium [5,6].

Copper is thermodynamically stable in oxygen free crystalline groundwaters; the long life expectancy of copper containers is based on this fact. The slight corrosion expected is due to residual oxygen in the repository and the production of other corrodants e.g. sulphide from microbial action [7].

Stress assisted cracking of these materials under repository conditions has also been investigated [2,6,8]. Although environmental assisted cracking of the candidate materials is thought unlikely, carbon steels (e.g. in carbonate solutions) and the titanium alloys (hydrogen embrittlent) may suffer stress assisted cracking in corrodants present in the ground-waters. It is not clear, however, whether conditions in a repository are sufficiently severe for environmentally induced cracking to occur.

Repositories in rock salt.

Rock salt formations are dry, but for the purposes of safety analyses the ingress of water as concentrated brines is postulated. Various brine solutions have been used in corrosion investigations for materials for rock salt repositories. The brines are either saturated sodium chloride with small additions of potassium, magnesium and calcium chloride or sulphate or magnesium chloride based such as Q-brine [9]:

$$1.4 \text{ wt\% } NaCl,$$
$$4.7 \text{ wt\% } KCl,$$
$$26.8 \text{ wt\% } MgCl_2$$
$$1.4 \text{ wt\% } MgSO_4.$$

Corrosion rates quoted for the lower pH magnesium brines are given in Table 5.

The general and pitting corrosion resistance of Ti-0.2%Pd and titanium grade 12 as well as Hastelloy C-4 are acceptable. The titanium alloys have been shown to be susceptible to crevice corrosion and Hastelloy to pitting under gamma radiation (oxidising conditions) [9,10].

Carbon steels are possible candidate materials as consumable containers provided they are not exposed to temperatures in excess of around 150°C. A container of this type is proposed for salt repositories [9].

Repositories in clay formations.

The containers will be placed in supported galleries in clay formations. These galleries are likely to collapse after the operational phase and the containers will be contacted by the clay or if a backfill is used by interstitial clay water. The composition of this water was [2]:

Water A: expected composition
13.9 g/l Na_2SO_4,
0.87 g/l K_2SO_4,
19.2 g/l $MgSO_4$,
3.6 g/l $CaSO_4$

Water B: a possible intrusion water
39.8 mg/l Na_2SO_4,
15.0 mg/l $MgSO_4$,
10.1 mg/l NaF,
74.2 mg/l Na_2CO_3,
39.4 mg/l KCl,
19.9 mg/l $NaCl$.

The corrosion rates measured at 90°C in waters A and B are given in table 6.

The results for clay and crystalline environments are similar; both Hastelloy C-4 and Ti-0.2%Pd were concluded to be suitable candidate materials; corrosion rates on carbon steels varied between 2 and 300 µm/yr depending mainly on the amount of oxygen available [2].

CONCLUSION

Overpack containers for high level waste with life expectancies from 300 to greater than 10^6 years appear feasible. There remains the problem of improving the database on these materials to improve the confidence levels in the life forecast. Pitting and stress assisted cracking, particularly the latter, are areas for which reliable data are lacking and are currently being investigated in more detail.

REFERENCES

1. Experiments on Container Materials for Swiss High-Level Waste Disposal Projects. Part III. NTB 86-25, NAGRA, Baden, 1987. J.P. Simpson. P.H. Valloton.

2. Corrosion behaviour of container materials for geological disposal of high level waste. Commission of the European Communities, Luxembourg 1986 EUR 10983 EN. Ed. B. Haijtink.

3. An Assessment of Carbon Steel Containers for Radioactive Waste Disposal. G.P. Marsh, K.J. Taylor, Corrosion Science, 28 (1988) p.289.

4. Development of Engineered Structural Barriers for Nuclear Waste Packages. PNL-SA-9543. Pacific Northwest Laboratory, Sept. 1981. R.E. Westerman, R.P. Elmore, S.G. Pitman, J.L. Nelson.

5. Waste Package Materials Screening and Selection, ONWI-312 Oct. 1981, Battelle, Columbus. Compiled by D.P. Moak

6. Long-Term Performance of Materials Used for High-Level Waste Packaging. NUREG/CR-4379 Vol.1. First Quarterly Report, Year 4, 1985. Battelle, Columbus. Compiled by D.Stahl, N.E. Miller.

7. Final storage of spent nuclear fuel - KBS 3. SKBF/KBS, Sweden.

8. Stress Corrosion Testing of Pure OFHC-Copper in Simulated Groundwater Conditions. Report YJT-84-21, Nuclear Waste Commission of the Finnish Power Companies. Nov. 1984. P. Aaltonen, H. Hänninen, M. Kemppainen.

9. Corrosion Behaviour of Container Materials for the Disposal of High-Level Waste in Rock Salt Formations. Commission of the European Communities, Luxembourg 1986 10040 EN. E. Smailos, W. Scharzkopf, R. Köster.

10. Waste Package Reference Conceptual Designs for a Repository in Salt. BMI/ONWI-517 Technical Report, Feb. 1986.

11. Corrosion of metals in Tropical Environments. Materials Performance, July 1976. C.R. Southwell, J.D. Bultman, A.L. Alexander.

12. Engineered Waste Package Conceptual Design, Spent fuel (Form I) Disposal in Salt. AESD-TME-3087, Mar. 81. Westinghouse.

5-20 years	Sealed in containers		
40 years	Intermediate storage		
ca. 40 years	Operation of repository.		
10^3 years ** to 10^6 years	Complete retention of waste within container		
10^3 years ** to 10^6 years	Slow release and migration of radionuclides.		

Table 1. Time scales involved in nuclear waste disposal and safety analyses. (** Times vary according to concept)

Geological Formation	Materials investigated 1 = Proposed container 2 = Under investigation	Life (years)	Country
Crystalline bedrock.	1. Cast carbon steel, Copper 2. Alumina	300-1000 300-1000	Switzerland
Crystalline bedrock	1. Copper 2. Alumina	10^5-10^6	Sweden
Crystalline bedrock	2. Ti grades 2 & 12, Copper Ni-Cr alloys (C-276, 625)	500 or 10^5+	Canada
Granite	2. Carbon steel, Hastelloy C-4 Titanium-0.2%Pd		UK
Granite	2. Copper		Finland
Basalt	2. Cast iron Ti grades 2 & 12, Ti-0.2%Pd	1000 min. 300	USA
Granite Salt domes Clay	2. Carbon steel Ni-Cr alloys (625, C-276, C-4) Titanium-0.2%Pd		France
Tuff	1. Stainless steel (316)	1000 min. 300	USA
Rock salt	1. Cast carbon steel 2. Titanium grades 2 & 12	1000 min.300	USA
Salt dome	2. Carbon steel, cast & wrought Hastelloy C-4, Ti-0.2%Pd Nodular cast iron, Si cast iron & Ni-resist	100+	West Germany
Boom clay	2. Stainless steels 304, 316, 904L, 1803 MoT, Ni-Cr alloys (C-4, 625) Titanium & Ti-0.2%Pd Carbon steel		Belgium
Boom clay	2. Carbon steel		Italy

Table 2. Geological formations and materials under investigation or proposed for HLW containers.

Water	mg/l	Na$^+$	K$^+$	Ca^{++}	Mg^{++}	CO$_3$$^-$	Cl$^-$	SO$_4$$^-$	F$^-$	pH
1 Basalt, synth.		263	2	1	–	97	148	109	37	9.9
2 Granite, synth.		106	–	20	6	244	36	24	2	9.4
3 Granite, synth.		4800	54	1100	3	–	8100	1820	4	8.5
4 Granite, max.		100	10	60	30	42	100	50	8	9.0
5 Sea water		11000	390	410	1300	140	19350	2710	1	8.1

Table 3. Compositions of crystalline rock groundwaters considered in some of the corrosion programmes.

Water		Material	General corrosion	Pitting	Ref
4	50-80°C	Copper	0.5 mm 10^6 yr	17mm 10^{6}yr	7
5	250°C	Ti gr.12	1.1 μm/yr (30ppb O$_2$)	see text	6
		Ti gr.12	0.6 μm/yr (500ppm O$_2$)	see text	
		Ti-0.2% Pd	1.1 μm/yr (30ppb O$_2$)	see text	
		Ti-0.2% Pd	0.6 μm/yr (500ppm O$_2$)	see text	
2	90°C	Cast steel	2.6 μm/yr (Note I)	see text	3
2	90°C	Cast steel	26 μm/yr (Note II)	see text	
3	80°C	Carbon steel	60 μm/yr (100ppb O$_2$)	(Note III)	1
3	140°C	Carbon steel	31 μm/yr (100ppb O$_2$)	"	
3	80°C	Carbon steel	10 μm/yr (2ppb O$_2$)	"	
3	140°C	Carbon steel	18 μm/yr (2ppb O$_2$)	"	
5	25°C	Carbon steel	65 μm/yr (Note IV)	158 μm/yr	11
5	122°C	Carbon steel	(Note V)	150 μm/yr	12
1	150°C	Cast iron	25 μm/yr (6 ppm O$_2$)	–	5

I. Specimens under 10 cm bentonite, solution open to air.
II. Specimens under 10 cm crushed granite, solution open to air.
III. No pitting observed, allowance 2 times corrosion rate.
IV. Natural sea water, well aerated, 16 years exposure.
V. Derived from Ref. 11. and short term high temperature data.

Table 4. Some corrosion data for candidate materials for HLW containers in crystalline rock repositories.

Material	Temp.	General Corrosion	Pitting
Ti-0.2% Pd	200°C	0.14 μm/yr	None at 6 months
Hastelloy C-4	200°C	0.21 μm/yr	None at 6 months
Cast steel	170°C	92 μm/yr	No deep pits
"	200°C	501 μm/yr	No deep pits
Carbon steel	70°C	70 μm/yr	
"	250°C	1700 μm/yr	

Table 5. Corrosion rates in anoxic magnesium chloride brines.

Material	Corrosion rate (μm/yr after 1000 h)	
	Water A	Water B
Ti-0.2%Pd	Weight gain	Weight gain
Hastelloy C-4	Weight gain	Weight gain
Carbon steel	214	148
C-steel welded	199	65

Table 6. Results of corrosion tests for clay repositories.